John Langdon Heaton, Frank Ver Beck

The book of Lies

John Langdon Heaton, Frank Ver Beck

The book of Lies

ISBN/EAN: 9783743324220

Manufactured in Europe, USA, Canada, Australia, Japa

Cover: Foto ©berggeist007 / pixelio.de

Manufactured and distributed by brebook publishing software
(www.brebook.com)

John Langdon Heaton, Frank Ver Beck

The book of Lies

THE BOOK OF LIES BY JOHN LANGDON HEATON

WITH MANY PICTURES
FROM PEN DRAWINGS
BY FRANK VERBECK

THE MORSE COMPANY
NEW YORK
1896

ARGUMENT.

To the lovers of truth these pages are commended.

Nothing is so truthful as an honest lie, since it bears its character writ large upon its frank countenance ; while truth, in the guise of incredibility, oft cries for credence like a hussy.

As for the lie insidious, the lie treacherous, the deceitful lie—sure 'tis only by study that we may learn to know it, as the surgeon probes the wound, yet unlovingly.

So here's to you readers, wishing the end of falsehoods and all manner of vain speaking!

THE AUTHOR.

CONTENTS.

THE BOOK OF LIES.

CHAPTER I.

"WHEN I WAS WITH JAMESON."

"WHEN I was with Jameson at Krugersdorp—"
began the returned miner—

"The mention of South Africa and 'the Crush-
ings of all the Rand' brings to my
mind," said Dr. Binninger, lighting
another of the returned miner's ci-
gars and hitching his chair a trifle
nearer to the big open fire, "thoughts
of the many curious ways in which
money is made and lost. Money
makes the mare go, and conversely,
the mare makes money go. The
horse, gentlemen—"

The boom of the doctor's big voice
was interrupted by brisk steps outside the door and the entrance of a
tall, lean man with a saturnine countenance and drooping black moustache, followed

9

rather hesitatingly by a short, fat man, with a close-clipped, pointed beard, who wore the air of a stranger. Together they suggested Don Quixote and Sancho Panza.

It was a strange scene they looked upon, as their eyes wandered over the dark walls and dusky ceiling of the quaintest club-room in New York. Heavy, sagging beams, black with age, spanned the low ceiling. The huge jaws of a fireplace, two hundred years old, were fanged with gigantic andirons curiously twisted out of wrought iron, and standing nearly man-high upon the uneven brick hearth. Outside, in what was by day one of the busiest downtown streets, reigned the quiet of the grave. It was as if the ghosts of all the dead and gone New Yorkers, who had two hundred years ago strolled Nassau street and the Maagde Paatje, had resumed their sway and driven forth the bustling presence of their successors.

Grouped round an open fire of big beech logs were about a dozen men, most of whom had the look of the typical New Yorker who has prospered, and whose appearance gives little hint of his profession.

The plump newcomer was introduced by his guide, long John Eckels, as Mr. Parker Adams, and was soon stretching his short legs toward the fire from a comfortable easy-chair in the wide semi-circle.

"As you came in, Mr. Adams," said Owen Langdon, one of the group of men by the fire, "we were speaking of horses, and I had it in mind to tell a little incident which once came to my notice. There was, about a year ago, a saddle-horse down in Bay Ridge which died of love, and for a lady. This animal cherished such a deep affection for its fair young owner, that it grew inconsolable when she became engaged to a bicycle young man, and gave up her daily canter in Prospect Park. Day after day the noble animal mourned in its luxurious box-stall, or moped about the pasture with dull eye and lifeless attitude. One day the thoughtless cause of so much misery came strolling by the pasture fence, leaning on the arm of the young man. Just opposite where the agonized animal stood, she took her lover affectionately by the ear, and, without apparent difficulty, dragged his head down upon her shoulder. Then, as she mur-

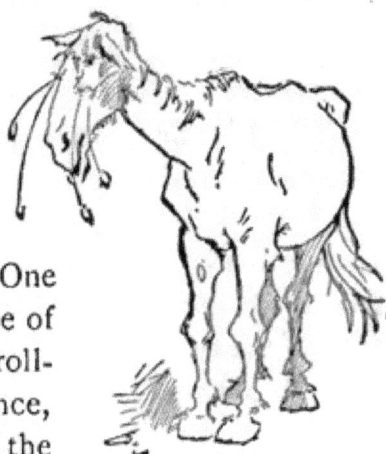

mured low in his ear, words of endearment, such as once she had lavished upon the horse, the intelligent beast's sufferings grew too great to bear. Waiting to hear no more than 'It's its own ownest darlingest's 'ittlest mopsy wopsy rosebud dumpling toad, so it is,' the maddened animal began sawing its throat along the barbed wire of the fence. The blood from its wounded neck flecked the fair cheek of the girl, and recalled her to her senses. By the most heroic exertions the life of the animal was saved, only that it might give, on a subsequent occasion, further proof of its devotion.

"The young lady soon forgot the incident and was married. A few weeks later—"

"Hem! The horse, gentlemen, as I was about to remark," said big Dr. Binninger, stroking his short, gray side-whiskers, "is a noble animal. It invented the horse-race when it gamboled with its mate in the unfenced fields of Eden, fresh from its Maker's hands. It was left for man to invent betting on horse-races. Man thinks himself wiser than the horse, but cannot eat so much, nor run so fast ; yet these are his best tricks. I am a Kentuckian, myself, as you know, hence I sympathize deeply with my old friend, Col. James Hudson, of Lexington, who, in the palmy days of the Hudson stables, was

one of the most open-hearted men of the old state.
But his money went up, paradoxical as it may seem,
in oil wells and silver mines, until a year ago all
that was left to him was one useless, old animal,
that he called 'Baldy'—short for Osbaldistone.

"Well, suh, one day, as Col. Hudson was lying
flat on his back in the pasture, meditating whether
it was unmanly for a Southern gentleman of the old
school to seek self-destruction, old Baldy came and
nosed him until he was compelled to get up, then,
gently gripping his sleeve, led him away to the
middle of the field, where he had pawed a deep
hole under an old oak. There Baldy went down on
his knees, and presently drew forth, with a whinny of
pleasure, a broad, golden coin, which he laid in the
colonel's lap. Gentlemen, it was a Spanish doubloon
of unimpeachable virtue and ancient mint. Hudson's gloom vanished in a minute. Running to
borrow a spade, he had presently unearthed a strong
box, heavy with similar gold pieces. Faithful
Baldy, unable to lift the box, had, with the stumps
of his broken, old teeth, painfully gnawed a hole
through the iron-bound cover, and taken out a sample
coin for his master. With tears of gratitude stream-
ing down his face, Hudson fell on Baldy's neck and
sobbed. He will never sell the dear old horse."

" Get much ?" queried Jim Hart.

" I never ventured to make intrusive inquiries," replied Dr. Binninger, rather stiffly ; "but common report placed the value of the contents of the box at $37,519. It is supposed that the coins were buried there by some of the early Spanish explorers of our noble waterways."

" Reminds me," said Hart, "of the hypnotic horse of the Woonasquatucket. Belonged to one Cargill, farmer down in Rhode Island. Went State Fair, Narragansett Park, last Fall. Remember ? Had a fakir there. Shammed dead. Hypnotic trance. Buried. Ten cents a peep. Horse got wind of it. Two days later Cargill saw horse in pasture, waving front hoofs before nose of susceptible young mare. Svengali business. Slow music. See ? By and by, young mare keeled right over into big hole, horses had pawed out with hoofs. Other horses just going to paw dirt back into hole on top of mare. Cargill said nay, nay. Always thought Cargill chump. Pity spoil experiment."

The fire crackled low, and presently, with a parting flicker, went out, leaving a great bed of glowing, coals that cast big, weird shadows upon the black walls. The room was dim with cigar smoke, and eloquent of peace, hospitality, and comfort.

Presently a deep sigh of content broke from the huge chair into whose depths John Eckels was luxuriously burrowing, and the knight of the sorrowful countenance spoke : " If it be truth the poet sings, that a sorrow's crown of sorrow is remembering happier things, then I owe you gratitude, Hart. The sense of having comfortably dined, the glorious fire and boon companionship give me pleasure's crown of pleasure in remembering a former condition which was next door to hell itself. Doubtless some of you have wondered how my cadaverous frame has been proof against the good cheer of the Travelers' Club, whose magic cooking is well advertised by the comfortable girth of some of you. I am sure no one could have guessed that my extreme emaciation is due to passing two years of my life in a hypnotic boarding-house.

" It was when I was a young man, newly come to the city," resumed Eckels, after a few quick puffs at his cigar, "that I visited, while looking for board, a certain house, which has since been pulled down. I didn't particularly like its looks, but for some reason, which I now understand better than I did then, after seeing the mistress, I felt irresistibly impelled to take a room. No, she was not personally attractive. I fared, as I then thought, on the fatness of the earth. So did we all. The guests reveled in roast turkey, in exquisite soups, in rare old cheese, and perfect coffee. Did anyone so much as mention a wish for a special article of food, it was forthcoming, with a smile, immediately or at the next meal. It ought to have been a happy family, yet the boarders somehow seemed a lean, ill-fed, and low-spirited lot, and, as you may well believe, none more so than myself.

"One day a tall, fine-looking gentleman called to look at rooms. If Mrs. Smith had known she was dealing with a professional hypnotist, she would have shown him the door. In a moment the professor felt that she was trying to hypnotize him to take a room. Exerting all his own psychic force, he soon had her will under complete control. He guessed the truth. She had for years been hypnotizing her servants to serve, and her guests to devour, slops for coffee, salt pork for tenderloin, and chuck roast for turkey ; and had amassed a considerable sum of money. She had not otherwise interfered with us, than to secure large rates for poor board. We were all able to attend to our work. She could have played hob with our business arrangements, and got rich faster, but I think she was afraid of criminal proceedings.

"Well, to make a long story short, Prof. Lauderdale was a good sort, and he resolved to rescue us, and feed us back to health at Mrs. Smith's expense.

"So he hypnotized the old lady in turn into pampering us with real luxuries, and into imagining that the codfish-and-prunes régime was still in force. Daily her ill-gotten gains melted away; daily the boarders regained their strength and spirits. Never was such a boarding-house known before or since! We breakfasted, lunched and dined like Lucullus. Mrs. Smith remained perfectly happy, little knowing that her hypnotic power was being sapped with her bank balance. Most of the boarders got as fat as butter, but in my own case, the lean habit was too firmly fixed to be eradicated. When we were all restored to pretty good condition, Lauderdale let us into the secret, and was amply rewarded by our gratitude. Mrs. Smith is now the unhappy inmate of a lunatic asylum, where she imagines that she is Queen Victoria, and dines daily on seven-course dinners, cooked by the fires of an ill-regulated imagination."

"When I was with Jameson at Krugersdorp—" began the returned miner—

"Just a moment, if you please," said Harry Porter, jumping to his feet, a tall, impetuous, fair-haired

young .ellow, looking like the scholar and athlete
that he was. "I see the night is getting old, and I
have an announcement that has been trembling on
my tongue all the evening. I would like that Jameson story saved till the next Ladies' Night, when—the
fact is, there is a—I mean," he ended desperately,
"I got examined for life insurance to-day. And—"

There was a moment of blank silence, when Eckels jumped to his feet, shouting : "Why, of course !
Boys, he's going to be married ! Give us your flip-
per, old chappie !" And in a moment Harry's
back was being slapped, his hand boisterously
shaken, his ribs punched, and a chorus of excited
comment arose.

"Break away !" shouted Tom Fenton, a man of
medium height, with a smooth-shaven, actor-like face,
and hair slightly tinged with gray," Give him a chance
to breathe. This mob and the mention of insurance remind me of a crowd I saw once in San Fran-
cisco. It was about 9 o'clock in the forenoon.
The throng reached right across Market street, and
tapered off away down the block. As luck would
have it, a runaway horse dashed into the thickest of
it, just as I reached the spot, and in the accident
seven insurance agents were slightly, and three seriously, hurt. About a dozen other insurance men

attended the wounded. This I learned afterward. A reporter of the *Standard* undertook to discover why so many insurance men were waiting on that particular corner so early in the day, and learned that there was within the building a man whose fire policy was due to expire at noon, and who thus far had declined to reinsure."

"I suppose, Harry," said Dr. Binninger, "that the young lady in the case is Miss Copeland, whom I have unfortunately seen only at a distance as yet. A charming lady, charming lady! Red-headed girls are the nicest—"

"It's the younger one; the brown-haired one," muttered Harry Porter, hastily, with a flush of annoyance.

"Just so," said the imperturbable Binninger, "let me congratulate you anew, sir! As I was saying, red-haired girls would be the nicest kind of girls, if it weren't for the incontrovertible fact that the other kinds of girls are just as nice as girls can be. Red hair, gentlemen, denotes an oratorical temperament. Fitzsimmons is red-headed. So was Thomas Jefferson. Cicero was red-headed, or else he wasn't. I have forgotten which, but really the point is immaterial.

"It seems undisputed that red hair generates heat

more rapidly than other kinds. I remember that
Mamie Wayne, a dear old friend I used to know in
Olathe, Kansas, was red-headed. One day she was
cleaning her last winter's blue silk with benzine,
when the inflammable stuff took fire, presumably
from her fair tresses falling low over her work. In
a moment the room was filled with heated air, which
expelled the cooler and heavier atmosphere. Kansas houses, I may parenthetically remark, are of very
light and flimsy construction, which partially ac-
counts for the easy victory they yield to the indi-
genous cyclone. As the air grew hotter from the
blazing fumes, the entire house, rising at first slowly,
soon shot upward like a Montgolfier balloon. There
was little wind, and it could presently be seen, at the
height of two miles, burning so furiously that not a
piece of it, except a few bricks out of the chimney,
ever reached the ground. My dear young friend
was never again seen alive. But come, gentlemen,
the hour is late, and some of us are no longer so
young as once we were."

With lingering farewells and bursts of laughter, the
men slowly drifted out of the room. There was the
sound, for a little space, of parting footsteps on the
deserted pavement, and then the narrow street com-
posed itself again to sleep until the dawn of a new day.

CHAPTER II.

"THERE IS NO 'ROAD REFORM' FOR THE FLINTY PATH OF TRUE LOVE."

"BEEN bicycling, hey?" queried Dr. Binninger, glancing down the dining-table in Owen Langdon's Brooklyn home. His eyes rested upon a pretty girl with brown hair, whom you would suspect at once as being the youngest of the Copeland sisters. After catching Harry Porter's look of devotion, as he gazed on her, you would have been sure of it.

"Yes," replied Mrs. Langdon, seeing that the lovers had not noticed the query; "alas for weary me, we have! If I ever undertake to chaperon another lot of young people on such a trip—well!"

"Roads good?"

"Fine!" said Harry; "we ran out along the Merrick road for miles. It's in perfect condition."

"I believe great progress has been made in road improvement in this vicinity," was Dr. Binninger's comment. "I regret that in my own beloved South a less satisfactory state of affairs exists, largely, it is true, owing to the unfortunate poverty of the people. Here and there, however, good work is being done,

as by my dear old friend, Dr. Onesimus Rylance of
Carson, Tennessee. Hem ! "

" Won't you tell us about it ? " queried the hostess;
" I'm sure it would interest our young wheelmen and
women."

" Hem ! " repeated the doctor, " if my explanation proves too technical for a social occasion, pray
check me. The lower orders of creation are, as we
naturalists know, less easily subject to death than
the mammalia. A sheep-tick's locomotive powers
are stimulated, rather than otherwise, by bisection,
and a decapitated snake will writhe, if not until sun-
down, at any rate a long time. This circumstance
suggested to Dr. Rylance the idea of splicing snakes
of different varieties to produce a less objectionable
composite reptile than most of those common in our
section. Selecting a rattlesnake and black-snake of
equal size, he decapitated both, and joined the rattler's body and the black-snake's head with a neat
suture. The result was a snake which could coil
and strike all day without inconvenience to itself or
others. Soon, in accordance with the well-known
Darwinian law of modification, the composite snake's
nose became indurated to a high degree. The doctor had merely intended to produce a curiosity, but
seeing how rapidly the compound snake gained

strength by exercise, he conceived the idea of usefully employing it. The result was so favorable, that he has now a whole detachment of Patent Compound-Snake Stone-Crushers at work, solving, for that part of Tennessee, the problem of good roads. It has been computed that a Compound Snake hits a stone with a force considerably greater than that of a hammer swung by an adult male human, listening for the dinner-horn."

Ann Copeland, who was in a mood to beam upon everyone, broke into a ringing peal of laughter.

"Oh, you funny man!" she gasped, bending upon the doctor a look from her big, dark eyes, which Harry considered a wasted sweet.

"My dear young lady," said Dr. Binninger, in his most grandiose manner, "I hope it is only because this is our first meeting that you find me amusing. In my own state, by men at least, I assure you I am accustomed to be taken seriously."

The offended doctor was a picture of pomposity Herculean of build, smug-faced, side-whiskered, he looked like a large paper edition of Depew, with wide margins, and the sense of humor expurgated.

"Oh, I say, Langdon," interposed Harry Porter, hastily, "what was the rest of that yarn you were telling us about the horse that fell in love?"

Langdon glanced around the table. There were present—besides the host and his wife, Dr. Binninger, and the newly-engaged pair—John Eckels, a young niece of the Langdons, as pretty as a picture, and as silent; and Miss Copeland, Ann's elder sister, of the Titianesque locks, and herself an exceedingly pretty girl.

" Oh, it wasn't much of a story," he began, modestly. " This horse, you must know, ladies, had fallen into despair because his young mistress, becoming engaged, had, with the cruel unconsciousness of love, neglected him. He attempted suicide by cutting his throat on a barbed-wire fence, but was rescued and healed. The wire was removed, and a plain board fence substituted. One day, however, the horse was missed from its pasture. The loss was promptly advertised. About three days later a handsome young Irishman came to the door —this was in Bay Ridge—in answer to the advertisement.

" ' And have you found dear old Hero ? ' asked the young lady, now a happy bride.

" ' Sure, Oi kem to tell ye, mum, seein' as its me day ahf, that day before yistiddy as I was goin' me roun's, jist in the gray o' the marnin'—Oi'm a poleeceman, mum—Oi see a moighty foine harse an-

swerin' to the description, a-standin' outside a build-
in in the foorst ward o' Long Island City, the same
bein' on me bate. Sure, it was a wise baste to get
that far, ahl by himself. He must ha' come ahl
the way 'roun' by Jamaiky an' Mashpet', bekase he
cud niver ha' crashed th' shwing bridge at Hunter's
Pint widout wings, it bein' ahlways as open as Tony
Miller's bar. So Oi kem to break it to ye, loike.
Ye'll niver luk upon 'im again in this loife, though
maybe in the hereafter.'

" ' But why? What was the building ? '

" ' Sure, mem, its meself's be the lasht to say a
wurrud to dim the eye of beauty wid one usheless
tear. 'Twas a sausage factory, and that's God's own
trut'.'

" Thus you see," concluded Langdon, " by the
exercise of what persistence and ingenuity Hero was
enabled to quit himself of a life which had grown
intolerable."

" Oh, how lovely ! " murmured Ann Copeland,
glancing shyly up at Harry; " why don't you ever
tell me things like that ? "

" Would you like me to ? " he murmured; " I'm
afraid I can't. I can only tell you the truth; you're
the dearest—"

" And yet," said John Eckels, pulling at his long,

drooping moustache, "I suppose dogs surpass all
other animals in their capacity for affection for human beings. I myself once owned a collie which
delighted in going with me on camping expeditions.
On one occasion a cold wave struck the camp, and I
began to cough badly. Just then I missed my dog, and
was in very bad humor about it for a whole day, but
at night was astounded to see the animal returning
at full speed, with a couple of thick flannel shirts
and a bottle of something pretty good for coughs,
gripped firmly in his teeth. He had traveled twenty
miles to get them from a country store, which was
left unwatched at dinner-time.

"On another occasion he saved me from a rattlesnake's deadly bite by thrusting his own body between me and the angry reptile. When bitten, off
he started at full speed. Thinking him mad, I ran
after him, and, following his trail, presently found
him in a moonshiners' camp, whose existence I had
never suspected, eagerly lapping up a little pool of
anti-toxic remedy which he had obtained by knock-
ing over and smashing several demijohns. He recovered all right.

"I think," Eckels went on, "he was the most
intelligent dog I ever knew. When I was living in
Chebunticook, Maine, he established in winter a to-

boggan slide of his own on one of the steep hills
that abound in that town. His first purely imitative
step was to tear a huge strip of birch bark from a
tree. The surface of this he rubbed with a cake of
soap, abstracted from my guest chamber. He used
to froth so at the mouth from chewing soap, by the
way, that he started a fine mad-dog scare in Chebunticook, although it was midwinter. His method
of using the toboggan was to lie down upon it at the
hilltop, holding the lower edge up with his teeth and
steering by dragging along the snow his bushy tail,
which, after three days of coasting, was worn as
bare as a rat's."

"Now you tell one, Harry," whispered Ann Copeland.

"I can't, love; on my life I can't," groaned
Harry, in visible dejection.

"Very well, Mr. Porter! Don't give yourself
any trouble on my account," she said; and before
the bewildered Harry could reply, the girl had
turned, with her sweetest smile, to Dr. Binninger, who
sat at her other side.

"Doctor," she began, "I am very much interested in the subject. Surely you, with your wide
knowledge of men and affairs, must know of similar
instances of canine devotion."

"Certainly, my dear young lady ! bless me, yes !" responded Dr. Binninger; and the vast smile, evoked by the flattering appeal, was marvelous to behold. "Hem ! If the subject is deemed one suited to a mixed gathering, I will relate a little reminiscence of a famous temperance revival in Yarmouth, Nova Scotia. Jerry Dibdin, a noted drunkard of the place, had a dog which became much interested in the meetings. Again and again the animal would run to the door of the little open tent, under which

they were held, whine wistfully, and trot away again. One day Dibdin, drunker than usual, was unsteadily standing beside a low hand-cart in the long village street, when Rover, who had never before been known to do so rude a thing, ran plump against him, upsetting him into the cart, where he lay sprawled upon a few remnants of—ah—of deteriorated fish, utterly unable to rise. The dog, with one quick, glad bark of joy, grasped the handle bar of the cart, and tugged away at it until he finally trotted, with his queerly assorted load, right down the middle aisle of the tent, and paused, panting, but happy, at the mourners' bench. When Dibdin recovered, he was so moved by this proof of the faithful animal's devotion, that he did not touch a drop of liquor for seventeen days."

" Oh, if I could only be loved like that ! " sighed Ann Copeland, gazing appealingly up at the gigantic narrator.

What Dr. Binninger said in reply was never known. It could not have been wise, or worthy of recalling.

" Fie, Dr. Binninger ! " was Ann's response, " if a man is as old as his heart, who here is so young as you ? "

" The little minx ! " thought Mrs. Langdon. But she only said: " Owen, what was the queer incident you were once telling about the dog that could count ? "

"Crows can count two, but not three or four," observed Harry Porter, in a desperate attempt to win laurels as a raconteur. Then he blushed and became six feet two of blonde embarrassment.

" The incident I am about to relate did not come

under my own eye," began Langdon; "a business acquaintance told me about it. For a long time the owner of the animal, Bent Burdock, a methodical old mountaineer in Murfreesboro', Tennessee, was puzzled to guess how the dog managed to know exactly at what hour in the morning to awaken him, by thrusting its cold muzzle against his face. One day he was permitted to oversleep himself two hours. Waking with a start, he presently remembered that on the previous night he had omitted to take his regular 9 o'clock dram of tansy bitters. The truth was out. The dog had simply reckoned from 9 P. M. to 7 A. M., eight hours, by counting the monotonous thump, thump, thump on the floor, as he wagged his tail, 72 thumps to the minute, 4,320 to the hour, waking Bent just after the 34,560th wag. The dog had learned to count by tending sheep."

"Do you see," said Mrs. Langdon, in a low tone, to John Eckels who sat at her right, "how that outrageous girl is making love to the old fool? There! I ought not to speak so about a guest! Owen would ask him and his silly old stories!"

"Owen is a pretty good teller of tales, himself," remarked Eckels.

"Yes, but that's different," said the lady, turning upon Eckels such a pair of clear gray eyes that he presently fell to thinking, with his divided mind partly upon Miss Copeland's ruddy locks, that he also would like to have such a champion.

" Please tell something, Mr. Eckels," added the hostess, in an undertone ; "anything to distract her attention. Look what a picture of misery poor Harry is !"

" Perhaps," began Eckels, hastily, "Miss Ann might be interested in a big Newfoundland dog I used to know in Paris, Maine. This dog, which was formerly accustomed to attend the Episcopal Church in that town with its owners, is one of the treasured possessions of the Bowman family, now residing in another village. In their present place of residence there is no church of the Episcopal denomination, and Bones, the dog, has often been coaxed to enter the Methodist Church, but has as often refused. One Sunday morning, not long ago, the family was horrified at missing the baby, baby-wagon and all, but was relieved, after hours of harrowing anxiety, by a telegram from Paris : 'Baby all right.' The dog had risen at night, wrapped the baby up warmly in its carriage, in which it always slept, and, taking the handle bar in his teeth, wheeled it eighteen miles to its old church, where he presented it at the altar-rail for baptism.

"Circumstances and personal taste have combined," Eckels went on without pausing, as if fearing an interruption by the big ex-Kentuckian, "to make me something of a connoisseur of dogs. I have known two who could infallibly foretell the weather. One belonged to Asa Ackerson of Manhattan, Kansas. Ackerson valued his dog at $3,500. For twenty-four hours before a rain storm, the animal will eat nothing but grass, refusing the most tempting ham and eggs, with milk gravy. Thus Ackerson has not only timely warning of storms himself, but makes a pot of money supplying phonographic weather warnings to 317 farmers, at twenty-five cents a warn, besides winning a good sum hiring the dog out to do lawn mowing in wet weather.

"The other dog was the property of Bently Whitman of Topeka, Kansas. Kansas has such a lot of weather that it develops the meteorological faculty in every one. Benjy—that's Whitman's dog—had only three legs, and was blind of one eye, but he was dearly loved. For twenty hours preceding a rain storm, Benjy carried his tail curled between his legs. When the weather was 'set fair,' the tail curled over his back jauntily, and if there was reason to apprehend high winds or a storm, he used to wag his tail excitedly, and lie near the door of the cyclone cellar. Some miscreant, one night, cut off the dog's eloquent tail. Benjy was overwhelmed with grief, and after a month of repining, during which time a number of disastrous weather changes came unheralded, finding that he could no longer be of service, he deliberately committed suicide by eating a number of railway restaurant sandwiches."

"Hem! I—"

Dr. Binninger was visibly expanding beneath the smiles of perverse Miss Ann, and seemed anxious again to claim the attention of the company, but, rising hastily, Mrs. Langdon led the ladies from the room.

She left behind one honest young heart clouded by profound gloom. She was followed by one young woman, already repentant of her cruelty, but resolved to continue it forever. For such, alas! is the nature of the sex miscalled, with fine irony, "the gentle,"

CHAPTER III.

LOVE WILL FIND A WAY? NO; INGENUITY!

"When I was with Jameson at Krugersdorp—"
began the returned miner—

"Sh-h! Not now!" whispered John Eckels,
hastily. "You were to tell that story Ladies' Night
for Harry Porter's girl, and now the engagement is
broken. The devil alone knows what's the matter,
and I'm going to find out."

Probably Eckels didn't mean to imply that he was
deep in Satan's confidence; for presently he had
manœuvred Harry into a dark and quiet corner of
the big meeting-room of the Travelers' Club, and
was asking how it all came about.

"Oh, I don't know, John, and I don't much care,"
said Harry, wearily. I suppose I, like an ass, was
angry because she flirted with old Binninger, and
then she said she was afraid she'd made a mistake;
that she never could admire mere muscle, but preferred the society of men of intellect. Oh, hang it
all! You know."

"She must think a great deal of you!" said
Eckels.

" Hell, yes ! heaps ! " said Harry, sardonically.

"Oh, yes she does ! You don't know women.
Nobody does, least of all themselves. I think I
know 'em, but it's only because I'm a bachelor and
never had any sisters. Oh, yes, she's utterly miser-
able, and fully persuaded she's happy and sure she
doesn't care for you, yet cherishes, all the while, a
sneaking expectation that you'll be suing for pardon—"

" But she's wholly in the wrong."

" My God ! Is it so bad as that ! Poor devil !
Then she'll never forgive you—never until, that is.
Strange creatures, yes. Every girl believes most of
the married women she knows are unhappy, yet she
is anxious to try the experiment herself, and thinks
she isn't. The girl that promises to be a sister to
you will be mad as thunder if you make love to another girl. You might try that. No? Couldn't?
Well, well ; don't worry because she's refused to
marry you ; that shows she intends to do it. You
reason by contraries, you know. A woman can always forgive a big offense easier than none at all.
You must insult her, boy ; scorn her ; make her
keep thinking of you, no matter how huffy she gets.
Measured by woman's scale, a quarrel and a kiss
average up better than indifference. Women are

heroic in crises.   A woman will scream at a mouse,
and spank a tiger with her broomstick.   Women
may average no higher than men in a general round
up of all the virtues ; but they're so strong where
we're so weak that we think 'em angels.   And by
George, they are ! "

" Didn't know you were so enthusiastic about the
sex, Eckels.   Isn't usual in an old bachelor like you."

" Old ?   Me ?   Why, you libelous young babe-
in-arms, you're twenty-six yourself—all of you except your intellect—and I'm only thirty-five.   I
suppose you think I'm a sort of Methuselah because
I'm not fat, but you wait !   When you're puffing
round at fifty years and 250 pounds, I'll still be as
slender and graceful as a gay gazelle.   No ; I'm a
bachelor all right.   I've had a hard fight against the
world, but things are coming my way now, and I
don't mind confessing that I am thinking about being your brother-in-law."

" The flaming locks have fired your old heart,
eh ?   Congratulations in order ?"

" Well, not just yet.   It will never do to appear
in too much of a hurry to close the bargain.   Keep
'em guessing is my way—in theory, you know, in
theory.   Since you're to be my brother-in-law, I
don't mind helping you out a bit.   You ought to

shine more in conversation. Beat old Binninger at
his own game. Confound a widower, anyhow!
Practice some anecdotes beforehand and tell 'em
with aplomb. Don't talk to her, but at her. Let
her see that others admire you, and she'll be proud
of you."

"You're a good fellow, Eckels," said Harry,
cheered in spite of himself by the other's nonsense,
"but I can't remember a story, and wouldn't have
cheek enough to tell it right if I did."

"Well, practice a few easy ones. Here, I'll give
you some of mine. There's a pretty good tale about
a wonderful surgical operation—the implanting of
a conscience into a traveling soap agent, you know.
Let's see : The operation which was performed at
the Johns Hopkins Medical School in Dahlonega,
involved a dissection of a lesion in the anterior med-
ullary surface of the unfortunate man's pineal gland,
and was so successful that the man, upon recovery,
became utterly unfit for his former business, and
has had to embark upon newspaper work. That's
only the outline, you know ; work up the ghastly
details in the dissecting room. Make 'em shiver.
Or—hold on—here's a better one : A serious dimi-
nution of Western Union Telegraph tolls is one result of the establishment of Pete Porter's bicycle

hen express between Alameda and Oakland—that's
the way you want to tell it.   Work in the long sen-
tences just as easy as if they came natural to you.

The towns are practically conterminous, but scatter
over a wide area.   Porter has a number of very intelligent hens trained to their work, for whom he

has fitted up small safety bicycles, scarcely more
than toys, weighing about a half pound each. Seated
firmly astride of her bike, a hen will, by combining
a vigorous flapping of her wings with strong pedal-
ing, get up very rapid motion on the level. On
down grades the movement is a combination of ped-
aling and flying. The towns are districted, and each
hen is domesticated in its own district, to which it
will return with the utmost speed, bearing whatever
message may be intrusted to it, which can then be
delivered within the district by hand. One of the
bicycle hens thinks nothing, when hurrying, of making seven miles in about five minutes, thirty-seven
seconds. Never let a lie come out in exact figures,
Harry."

"Now, see here, Eckels, what's the use of talking
nonsense? I can't remember such a string of lingo
as that. Try me on some easy ones, and I'll tell
Goldilocks you're a capital fellow."

"For God's sake, don't! Hint to her that there
is a sad mystery in my past that has blighted ten
of the best years of my life. Tell her you've heard
I'm the victim of deep-rooted sorrow for a woman
unworthy of me. But don't call me a good fellow!"

"See here, Eckels, get down to the lies."

"All right. Here's a pencil. Now take notes.

That's a pretty fair lie about a gang of laborers, near Nashville, Tennessee, repairing a bridge, when a black and sprawling object loomed suddenly in the sky and plumped down into the water before their eyes. Presently a man's head rose to the surface. They pulled the stranger out, when, puffing and blowing the water from his mouth, he said : ' Morning, gents ! Rather breezy to-day. Will any of ye be kind enough to tell me how fur I am from Kansas ? ' ' Cyclone ? ' inquired a workman. ' Oh, no ; not a cyclone, exactly ; lived in Kansas ten year and never saw a cyclone ; just a bit of high wind.'

" Here's another: A man in Manitowetiwockelagunticookisametic—oh, any old name will do for Maine ; take a day off to put in more syllables— has a tame seal, whose intellect has been developed to a remarkable extent. The seal's owner is a salt-water fisherman, and always carries the pet with him in his little schooner. Arrived upon the fishing-grounds, the seal slides overboard, and, guided by unerring instinct, locates the schools of fish and leads its master to them by coming to the surface, slapping the water with its tail and emitting short, sharp barks of joy. When the fisherman has secured a load, he heaves to and blows a big dinner horn, and the seal comes aboard for its rations of bread and milk. Then it curls up by its master's side, and goes to sleep in perfect contentment.

"Something like that is the story about the smuggling steer. Put it down like this : On the Ohio side of Lake Erie, at its widest point, is a deserted strip of shore untended by the Custom-house men, where, until the trick was recently detected, a very curious fraud upon the revenue was carried on by Andy Bogart and his trained steer. It was Andy's custom to go over to Canada, buy a lot of likely young cattle, and drive them with his steer to the lake edge, when the steer would swim across the lake, leading his convoy over. The ruse was discovered by accident, through a lake skipper running upon a fine herd in transit. At first he thought it a shoal of new-fangled horned porpoises, but when he discovered the truth he gave information to the officials at Cleveland. You could put some very fine descriptive touches into that."

"No, I couldn't either Jack, and you know it. I could learn it like a recitation, but I get so blamed red in the face."

"I see. Got to have 'em shorter. Kind of thrown off careless, like. Try a highly condensed lie, delivered in a bored and languid tone. This way : 'Ya-as, like the gas tree—aw—down West Virginia, you know ; soil so—aw—impregnated with gas they set fire to trees, and—aw—light up surrounding scenery—aw—so that bees can work all summer twenty-four hours a day, preparing for the night that never—aw—comes.' There's a good short lie about rats in the mines stealing the miners' lamp oil by dipping their tails in it and licking them. Let's see ; how did they unscrew the cover ? Oh, I remember: the biggest rat grabbed it in his teeth and the others took his legs and swung him round like a capstan bar—see—till the cover came off. Story of a naughty dog locked in house ; gets umbrella ; goes up to roof ; holds handle in mouth ; jumps off ; parachute descent and consequent forgiveness.   Story of—"

"Hold on ! Hold on !" cried Harry. Eckels was gesticulating rapidly, his thin face having passed from half jest to wholly earnest.

"I shouldn't wonder, though," said Eckels, "if aphorisms were more in your line. A touch of cynicism goes well with women ; makes 'em think you're a sad dog and have lived and suffered ; then they want to comfort you and make your paths all peace; only they wouldn't be, you know. Something like this : 'One's enemies pity him when he fails ; his friends blame him ; his family forgives him after many years.' ''

" I could do that better," said Harry, judicially. " It's shorter, and any fool thing seems to answer for an aphorism. If a dollar bill is the square root of all evil, a gold brick is its cube root. If wishes were horses, and were also father to the thought, then the thought would be a colt. Mountains have to keep pretty still because they are confined in mountain chains. A fiddle is masculine because the most successful appeal is always made to its stomach."

" Oh, you young idiot ! You'd spoil it all. You should be the cynic, not the clown, to impress women, especially a goose of a girl. I give you up. No, by Jove ! I have it !" said Eckels, jumping to his feet. " Will you trust yourself absolutely to me, upon a positive guarantee that my plan will restore you to happiness ? "

" Yes, I'll do that. Probably your scheme is a foolish one, but in love the most foolish thing is wisest.''

" Good boy ! That was an aphorism worthy of one whose name modesty forbids me to utter. Talk like that—but not to your dear one—to other women. For my first command is that you make no effort to see or speak to Ann Copeland until I give you leave, save the most ordinary courtesies when you meet by accident. Mind, now ; not one word about love or sentiment, and appear as jolly as a lark. Trust me. I'll see you through all right."

CHAPTER IV.

LOVE LAUGHS AT LIES.

ECKELS had made the acquaintance of Parker Adams at his boarding-house, and had secured his election as a member of the Travelers' Club, on the representation that Adams was himself something of a concocter of travelers' tales.

A burst of laughter from the group at the fireside proved that Adams was triumphantly standing his initiation. As Eckels and Harry Porter neared the group, he was talking about chickens.

"The hen," said the pudgy little fellow, on whose bearded face rested a look of deep gravity, "has wonderful vitality, and is for this reason capable of undergoing strange surgical experiments. At the boarding-house where Eckels and I are staying, we have an especial breed of four-legged chickens from the farm of Ex-Gov. McCarty, of New Jersey, who is an enthusiastic fancier. The first quadrupedal chick McCarty ever developed, by grafting the additional members, is alive yet, and is a methodical old bird. It uses only one pair of legs at a time, curling the spare set up under its body like a stork. Every half hour this rooster changes legs, and such

45

is the unerring accuracy with which this operation is timed that the crew on an accommodation train, which stops at a station near by, are accustomed to set their watches by the four-legged fowl's change-off.

" I have a friend in Rome, Georgia," Adams went on, " who conceived an idea fraught with woe to his snake neighbors. He selected for the experiment a chick that had been half swallowed by a snake when very young, and had lost one leg in the mêlée. For this valuable fowl my friend rigged up a wooden leg. Ever since then the bird has taken revenge on snakedom. Its method, when it sees one of its enemies, is to advance the wooden leg cautiously. The snake bites into it, and is unable to release its fangs before the chicken picks its eyes out. The rest is easy. The wooden leg has to be renewed twice a week, it gets splintered up so badly.

" Another instance of the vitality of the hen," continued the new member, " was brought to my notice while I was fishing in Maine last year. A farmer, one year previously, had accidentally overturned a water-bucket, imprisoning a hen beneath it. A lot of hay was subsequently piled upon the pail, as the barn was filled, and it was only removed a year later. As the hired man lifted the pail, the

hen ran away, rather groggily, but defiant, sounding
her alarm.   She had not been idle during her long
imprisonment without food or air, but had laid nearly
a bushel and a half of eggs, the sort known as
' Stage Elevator Ds.' "

"A most interesting occurrence!" said Tom
Fenton, his keen black eyes sparkling.   "I think
that we have reason to congratulate ourselves upon
our new member. He will be an honor to the Liars'
Club, to give our organization its popular name.   I
will now, with your permission, relate two or three
little instances which go to prove alike the folly and
the faithful devotion of hens to their duty. One case
is that of Harry Tyrwhitt, a shrewd farmer of Acco-
mac, Maryland. This gentleman, in order to induce
greater activity of laying among his 120 hens, has
formed a championship club league of twelve clubs,
each club containing ten hens.   Match lays are laid
by the clubs, two by two in rotation, and the score
of each club, from the first to the last day of the season, is kept with the utmost accuracy.   Under the
wholesome spur of emulation there are never any
goose eggs.

"Of course, the strain of this competition is tough
on the hens, but nothing like so cruel, after all, as
the course pursued by a hen-farmer I used to know

in Scanderby, Michigan. In spite of the hard times, this man has been gaining wealth so rapidly by the sale of an enormous number of eggs, without any visible supply of hens to account for them, as to suggest to his neighbors the idea of witchcraft. A committee visited his barn one rainy night, and there discovered that he had built a dark room, wherein were impounded twenty-six weary, pallid hens. The walls were painted to imitate green fields and running brooks. A powerful electric light shone for half an hour, and then for an equal time all was dark. The hens under this treatment had lost their reckoning, and were producing each twenty-four eggs per day."

"I must take issue with you, however, on the subject of the hen's intellect," said the new member, stroking his short, pointed beard. "Probably you know that in Ashaway, Rhode Island, reside a great number of Seventh-day Baptists, who are all very good people and religious almost to a fault. Among their other peculiarities is the possession of a special breed of hens, carefully trained to lay eggs every day of the week but Saturday. One of these hens was recently sold to a new owner in Chepachet, whose other fowl were irreligiously accustomed to lay eggs on all days of the week. The new-comer, upon fully realizing the character of her associates, began destroying the works of the devil, by breaking with her beak every egg laid on Saturday. This practice soon rendered the pious hen so unpopular with the farmer's wife that she met the fate of a true reformer, and lost her head. You will hardly dispute the statement that this hen proved herself the possessor of a high order of intellect, as well as a thoroughly developed moral sense."

" To be beautiful is the most useful thing a woman can accomplish," said Harry Porter, unconsciously speaking aloud an aphorism he had been busily concocting for the past hour or so. Then, catching the wondering looks of his companions, he was silent again.

" Hem ! As the subject has been somewhat abruptly changed by our young friend," said Dr. Binninger, "I think I will tell of a peculiarity of Southwestern Missouri. The soil of that section, as you doubtless know, is so strongly impregnated with iron that it produces a peculiar effect upon pigs. It doesn't injuriously attack their hoofs, but turns their nostrils into iron, so that double-barreled Fourth of July salutes can be fired from them without inconvenience to the pig.

" My very good friend Boanerges Smith, with

whom I was once staying, had a pig whose tail he
had neglected to cut off, according to the usual custom of the country.   In time the iron of the soil
collected upon the pig's tail in a hard and heavy
mass, which could not be got off except by smelting,
and that painful process would have scorched the
pig's coat tails.   The constant strain on the animal's
eyes, due to its inability to wink because of the
weight of the tail drawing its skin back, produced
partial blindness.   Smith provided a pair of blue
spectacles and endeavored to induce the pig to wear
them, but the maddened animal would insist on paw-
ing them off.

"At last my friend resolved to do that which he
should have done at first, and cut off the animal's
tail.   But it was too late.   The pig, relieved from
the load which had so long cumbered it, testified to
its joy by capering about.   But a new difficulty
arose.   The hind legs, by carrying such a weight,
had grown so much stronger than the fore legs, and
outran them so far and fast, that the pig went round
and round in a narrowing circle."

Here Dr. Binninger paused, and began leisurely
puffing at his cigar.

" But what became of the pig?" asked Porter.

" If the club will excuse me, I'd rather not tell.

It seems to me that even the uncompleted narrative has a certain modicum of scientific interest," said the doctor, and to all urging he remained obdurate.

" The voracious appetite of pigs is one of the most wonderful of their many admirable qualities," said Fenton, after a pause. "Sometimes this gets a pig into trouble. It was in Osgood, Indiana, that Killian Dopp's pig got into a hardware store and ate a lot of nitro-glycerine. Happening afterward into Raw's livery a horse kicked poor piggy. The stable was only slightly damaged, but the pig was blown to fragments, and some of its bristles were driven through a three-inch oak plank ; and the atmosphere for miles around was so greased with lard that it slipped into a cyclone. In the South I used to see sometimes a racing pig belonging to Bud Davis, of Mobile. It is an eighth wonder. This pig, which is kept lean and in good condition by a diet of pick-led olives and lean beef, with celery and coleslaw as side dishes, will drag a tiny sulky at a 2.40 pace. The only difficulty with him is that whenever you wish him to go east you have to start him north-west by the compass."

"In the face of such instances, what a mockery is the fancied superiority of man !" mused Eckels. "A man has little eyes to see with, middle-sized

arms to work with, and great big legs to gad about
on.   He gets most honor when he is least a man ;
when he has become short-sighted, fat, scant of
breath, bald, timid, feeble and a fool.   I like better
to think of primal man, of the Aborigine.   Such
thoughts were forcibly borne in upon me by the
recent discovery, in Bucyrus, Ohio, of a tomahawk
buried deep in the heart of a giant tree.   This imple-
ment differs in form from any now known, and has
deeply graven in its handle ' H. to M., B. C. 1317.'
It is supposed to have been given in the year named
by Hiawatha to Minnehaha, upon whose fair young
life the gift of an edge tool naturally cast the dark
shadow of an unappeasable hoodoo."

  "A curiously similar incident," said the new member, " recently came to light in a Virginia forest, when
a very large tree was cut down, near the heart of
which was found a lock of hair, or rather two, inter-
twined, of darker and lighter meshes.   By carefully
splitting the wood, the initials ' I. S.' and ' P. P.'
were discovered, faintly legible, in what must have
been the bark several hundred years ago.   By carefully counting the rings which covered the initials,
they were found to support the theory that the mys-
tic letters stand for ' I-o-h-n,' or John Smith,' and
' Pocahontas Powhatan,' whose locks of hair had been,

in some romantic mood, intrusted to the keeping of the cleft bark. Smith, you know, was one of these light-complected cusses, and Miss Powhatan was, of course, an equally pronounced brunette."

"The starry heavens and the mind of man," resumed Eckels, "won Kant's wonder. But Kant was a man. I, myself, most admire animals when they imitate the ingenuity of man, and become thereby less like their own noble simplicity. I was once, while in Central Africa with de Brazza, immensely amused by seeing, near the Congo, a party of monkeys listening to a simian visitor with a brass collar, evidently an escaped pet. Presently all sprang up and, chattering about the explanation, began to play 'keeping store.' The other monkeys brought bananas and cocoanuts to the brass-collared one, who gravely paid for each purchase with a pebble. Then the monkeys wished to buy their goods back for the same pebbles. The ringed monkey had failed, or the money was no longer legal tender. At any rate the refund was refused. There was a whispered consultation, and the merchant, all at once, found himself kicking at the end of a Molopea Kalostoma vine thrust through his collar. It was a first rate example of the rude justice of lynch law."

"The fact that even insects are not altogether

without rules of concerted action," said Dr. Binninger, "has been noted by Lubbock and other scientific observers. The other evening in Waycross, Georgia, along line of lights was seen flying across a low, swampy piece of ground. Some curious observers drew quietly near, and saw that it was a procession of lightning-bugs, proceeding in orderly fashion by fours front, and with the proper interval. Their combined light was sufficient to make their motions perfectly visible. After passing through all the principal streets of the swamp, the procession halted in a clump of alder bushes, where a committee was hard at work preparing a collation. A giant lightning-bug sat on a mullen leaf, and as each platoon passed him the paraders flashed their lights three times in quick succession as a salute, the big lightning-bug flashing once in reply. After all had passed the collation of sorrel leaves was served. It was the opinion of the observers that the lightning-bugs were holding a ratification meeting in honor of their ruler for the season."

"A friend of mine in Colorado, a member of a local club, vouches for the truth of this incident," continued the Doctor. "Two hunters, coming to the edge of a glade, saw sixty-five bears in consultation. One was lying bound with grapevines, two

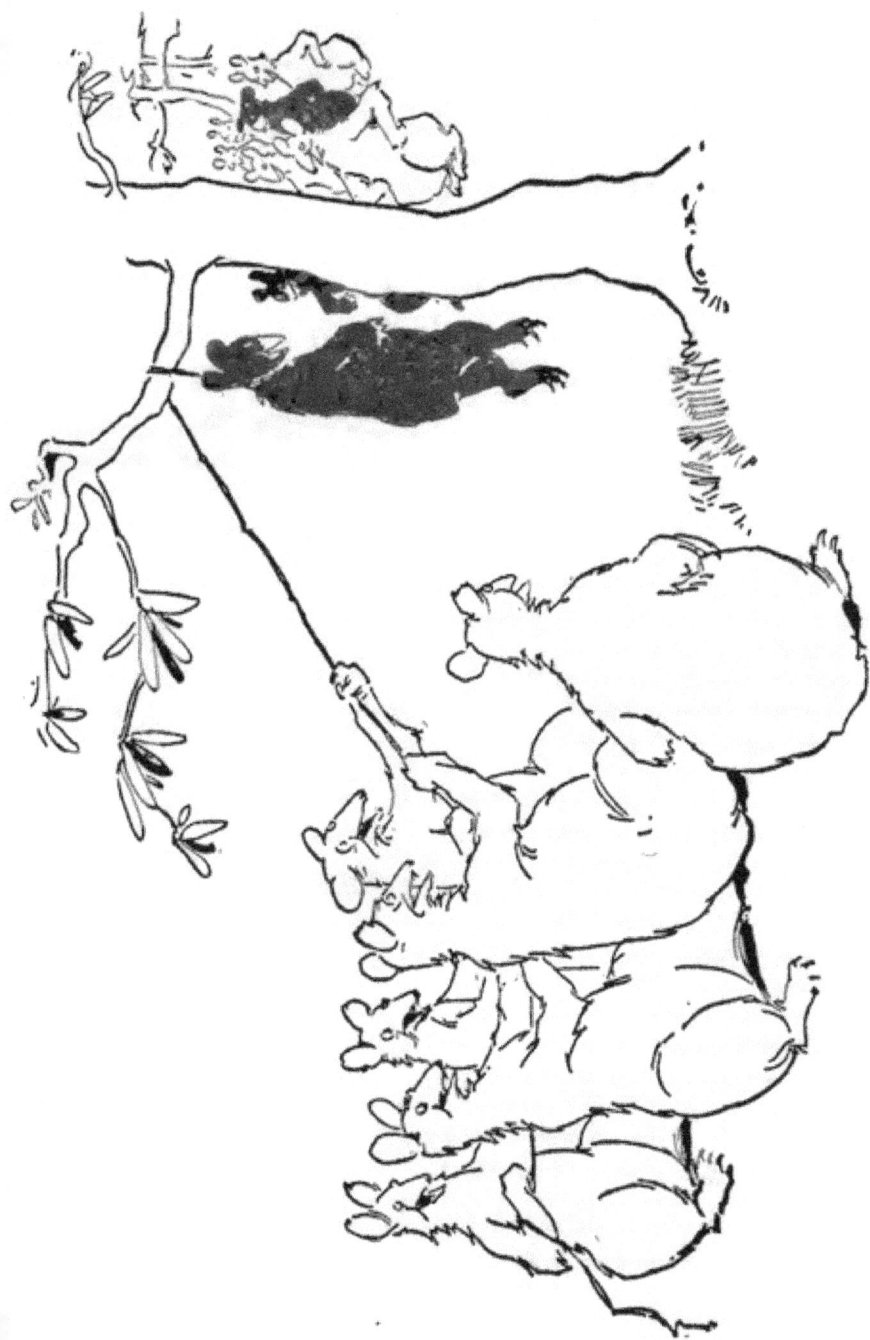

were watching over him, and another, a big fellow, was making a speech. Presently the big bear stopped talking, and all the sixty-three growled an assent. It sounded, my friend writes, like the muttering of distant thunder. The last sad act soon followed. Some of the bears threw a grapevine-loop about the erring bear's neck, led him to a limb, threw the vine over it, and six big bears walked away with it. In ten minutes the bear was dead, and the others went solemnly away. The hunters secured the lynched bear's skin without a bullet-hole in it, by way of proof."

"Was it ever known what the bear's offense was?" queried Fenton, who stood in his favorite attitude, both hands in his trousers-pockets, and his head thrown well back.

"No, but it is supposed that the erring bear had ventured the hasty opinion that honey belonged to the bees who made it—an opinion, as all must perceive, utterly subversive of morality and good government."

"The cat," said Eckels, "though seldom imitative of man, or influenced by devotion to him, often evinces reasoning faculties of a high order. Yet, however far toward reason, the instinct of cats may carry them, there is apt to be a flaw in the chain when they endeavor to measure and forecast the probable actions of human beings. Like the cat in Biddeford, Maine, whose owners recently removed from one house to another more commodious. Kitty was ill at ease, because she loved the old home, but no one attributed to her two or three mysterious fires which occurred in the new house, and were successively extinguished, until one day she was seen to leap upon the kitchen table, stretch up, and take a match from the tin match-box on the wall, and sneak away with it. She was followed, and carefully watched, so as not to excite her suspicion. First, she prepared a soft wad of old newspapers under the cellar stairs; she rubbed the match on the cement floor until it was in flames. Carefully pussy touched it to the paper, and the bright flames shot up, only to be again extinguished. Pussy was, after this exploit, killed, but with many regrets, for the family realized that her act was not malicious, but was intended merely to force them back into their old home by destroying the new."

"I think," said Dr. Binninger, "that the time has come to retire, for me, at least; and I, therefore, wish you, gentlemen, a very good night."

As the big doctor was leaving the room, the new waiter, Henry, nervously touched his arm. Henry

was a light mulatto, with a very earnest face, now
tense with excitement.   As the other members, one
by one, drifted out of the room, he said :   " 'Scuse
me, suh, but would yo' be kin' enough to tell me,
suh, wut become ob de ossifer pig ? "

" The ossified pig ?   Certainly, George, certainly.
I am always glad to see one in your station of life
eager to absorb scientific knowledge.   I may say
that I refrained from concluding the tale in open
meeting, only because I feared that some of the
members might assume an expression of incredulity;
and that, as you know, a Southern gentleman of the
old school could not tamely endure.   The ossified
pig, you'll remember, Charles, was left whirling
about in its joy at its release from an incumbrance,
and rotating more rapidly because of the greater
strength of its hind legs outrunning the fore legs.
Faster it sped; smaller grew the circuit of its revolution.   Presently blue smoke began to arise from
the no longer distinguishable mass of revolving
flesh.   Centripetal attraction was proving too strong.
There was a sudden flash, or report, an exquisite
scent of roasting meat—and the unfortunate pig had
vanished into thin air."

With a howl of anguish, the new waiter fled from
the room.

He tore up the stairs to a sort of loft he inhabited, with the club's other employés, and began hastily throwing his few belongings into his carpet-bag, and as he worked, he moaned, "Oh, wha' fo' dis niggah make sich a fool quesh'n? Oh, Lordy, Lordy, don' let 'im git me! Don' let 'im git me!'

"Wonder what ails that boy!" thought Dr. Binninger, as he slowly left the room.

CHAPTER V.

"DEAH me, what extwao'dinawy cold weather!" said the English visitor, after he had been duly introduced, and had drawn his chair close to the fire.

"We have, indeed, a variable climate," said John Eckels, "but you hardly get a fair sample here in New York; still, even here, during the '88 blizzard, a mouse crawled into my bed to keep from freezing, but wasn't quite quick enough; the last hind leg mousie pulled in, was nipped solid by the frost. Next morning the mouse ran and crouched over the register, and I then noticed that it had been turned entirely white by its awful experience; all except the frozen leg, which was black.

"During the same frightful storm, a Mrs. Alvan Learby, of Port Jefferson, was doing her washing. The room was, as is usual on such occasions, full of steam, and when Mrs. Learby's little boy, Jake, came running in, leaving the door wide open, the inrushing cold air condensed the steam into snow, which fell like an avalanche upon poor Mrs. Learby, completely burying her from sight. But Jake, by a desperate effort, succeeded in digging her out with a coal-scoop before death ensued from suffocation."

" Fawncy ! " ejaculated the visitor.

" At about the same time," went on Eckels, gravely, " a man was going up Broadway, carrying a good-sized jag—"

" W-W-Wait a minute ! " said his lordship, stammering in his anxiety, and tugging a big notebook out of his breast pocket. He was an angular man, with longish mutton-chop whiskers, a placid stare, and a monocle. Probably under the impression that to wear a dress suit would have been displaying too much solicitude for his appearance in an assembly likely to be uncouth in its attire, he wore shabby checks of very loud pattern. Jim Hart had brought him, and it was his first day in the country.

When the noble critic of our customs had secured his notebook and pencil, he said: " Mind my taking notes ? Want to put down 'jag means umbwella,' before I forget it. How vewy owiginal ! Thanks, my deah fellow ! By the way, 'jag' does mean ' umbwella,' doesn't it ? "

" Oh, of course," said Eckels. " Yes, ' jag ' does mean something to keep water out. I have forgotten what I was about to say, but as Mr. Hart tells me you are gathering material for a book, I will tell you about the effect extreme cold has on some of our species of snakes. Doesn't kill 'em; merely renders 'em lethargic. Out in Colorado, for instance, a Cincinnati man was building a sawmill in freezing weather, and having occasion to turn out a bar five feet long, ran out and got a long stick for the purpose. He cut off the small end to the right length, and put it on the lathe, when it began to thaw out, and proved to be a mighty big rattlesnake, recently deprived of its tail. But he never lost his presence of mind. Before the snake could coil for a spring, he plunged it in a tub of ice-cold water, after which it was easily despatched."

" Fawncy ! " said the visitor, still diligently writing, when he was not engaged in replacing his monocle to gaze at his informant.

" Yes, indeed," said Tom Fenton. " It is well known that snakes are frequently found in cold weather, frozen stiff, and so brittle that they can be broken like dry sticks. A snake so broken will rejoin himself as soon as warm weather thaws him out. Tom Norton, of Altoona, a few years ago, when I was living there, broke up several snakes, and piled up about half a cord, in stove lengths, of rattlers, moccasins, and garter snakes. A sudden thaw sent them hunting up their lost lengths, which all succeeded in finding but two, a big rattler head, in the confusion, joining a black-snake body, which, of course, left nothing for the black-snake head and rattler tail but to follow the example. Then followed a strange scene. The rattler body persisted in coiling up for a spring, when its black-snake head was unused to that method of fighting, and the rattler head and black-snake body were equally at odds ; so that neither combination could harm the other, though both were very angry. Finally the rattler-headed combination marched to the creek and committed suicide. The black-snake head would have followed the example, but the coiled body refused to move, and the composite snake soon died of a broken heart.''

" To meddle with snakes while in this lethargic condition is very dangerous,'' said Dr. Binninger. " In the Punxsutawney Mountains a farmer, named

Morris, once found a den of sixty-seven rattle-snakes, all rendered torpid, so that they could be handled with little danger. He prepared a big box, with holes in the sides for ventilation, corded the snakes up in it, and with his ox sled hauled it to a warm place. His intention was to sell the snakes for museum purposes, but when they awoke in their narrow quarters a terrific fight ensued, in which sixty-six of the snakes were killed, and the remaining one so crazed by pain, that he committed suicide by striking his fangs deep into his own body."

"Fawncy !" said the visitor, still writing.

"Cold weather produces some marvelous effects in states usually considered warm," continued the Southerner. "In the cattle belt of Texas a great herd gathered upon a railroad track at a point near Sherman, one very cold day last winter, and compelled the engineer of a freight to halt for fear of being ditched. After a fierce struggle, during which they used their long horns on each other with telling effect, the cattle huddled about the warm boiler to protect themselves against the cold, which was severe, until the fireman, as a last resort, drew the fire. After the cattle had deserted the fast-cooling boiler in disgust, he fired up again, and the train proceeded."

"I suppose bears hibernate," said the visitor, pausing for a reply, with poised pencil. "I am very much interested in bears."

"Hem!" said the doctor, "that bears frequently hibernate is true. That they always do so, may be disputed, in view of the experience of my friend, Maitland Brown, of Athlone, Oregon, who met a fierce grizzly prowling about the mountain, on one of the bitterest cold days of the winter of '78. The ravenous animal would undoubtedly have made a meal of Brown, who was unarmed, had he not retained presence of mind sufficient to spit deliberately, first in one and then the other of the animal's eyes. So extreme was the cold that they instantly froze shut, and the poor beast was like a blinded Samson. Brown then threw snow in the animal's mouth until, first melting and then freezing, the expansive power of the ice pried his great jaws apart, and the bear died where he lay."

"Fawncy! Such presence of mind!" ejaculated the visitor in amazement. "Bears must be very dangerous creatures, I judge."

"Troublesome rather than dangerous," said Eckels, "as a rule. I once knew a farmer who suffered severe loss because of the desperation caused among the animals living near him by cold weather. His house stands in a lonely spot, and he attributes to this fact his recent loss of a lot of farm stock. He had locked up his barn securely, but had omitted to fasten a second - story window, and during the night, three bears of assorted sizes, after trying all the other windows and doors, cast longing eyes at this. So the middle-sized bear climbed on the big bear's back, and the little bear, upon the middle-sized one, was just able to reach the window and open it. Then he went down stairs, unhooked the door, and the whole bear family came in, and ate, uninterrupted, veal and potatoes. When my friend, the farmer, came in the morning he noticed that the bears were

of polite ways, as they had used a number of empty grain bags for napkins, and a bran-mash tub for a finger-bowl."

" Hem ! " said Dr. Binninger. " The longer one lives and the more one learns, the better he realizes the truth of the central idea of Darwinism—the theory of modification by use and environment. Since the cold storage method of preserving eggs began to be practiced in New York, large polar rats, covered with very thick fur, have gradually come to inhabit the warehouses, making a living by gnawing the eggs. That these rats are not a special breed, but a variant from the ordinary species, is highly probable. Certain it is that an ordinary domestic cat, which has been playing about one of the warehouses for some years, has become so inured to the cold that she prefers the freezing air of the vaults to the outer sunshine. And just as the usual cat curls up to a grate or register in winter, this extraordinary animal, after a short run out-of-doors in summer, will return, with every sign of discomfort, to her vaults, and cuddle close to the genial surface of a block of ice. Her hair has grown very thick and long, and her appetite for cold-storage rats is voracious in the extreme."

" When I was with Jameson at Krugersdorp—" began the returned miner—

Just at this moment a diversion was created by Harry Porter dashing into the room, with a queer, strained expression in his eyes, and a generally un-cared-for look.

"Here, Eckels," he said, taking no notice of the others. "Can I speak with you just a moment?"

"See here, old man," he added, dropping his voice as the two moved away from the others, "I came to take back my promise. I must see her; I am desperate."

"How about her?" asked Eckels.

"Oh, her! She's happy enough, I guess. Gay as a lark. Flirting desperately with fellows that—"

"Good!" said Eckels. "The scheme is working splendidly. My boy, she is pining for you! Just you keep a stiff upper lip, and trust to me. I know something about women, I guess. Just look at me! I'm as much in love as you are, and going to be your brother-in-law. Do you see me going around with my necktie on crooked and a 'who-kicked-me?' look in my beautiful brown eyes? Not much! I'm here listening to yarns and letting her do the worry-ing. By the way, Hart brought in an English lord or something, and they're giving him points for his book on America. It's great! Come over and hear them. Do you good."

When the dissimilar pair of lovers reached the fireside group, Dr. Binninger was giving the dramatic conclusion of a tale about a monster flight of locusts, a dash of rain cloud, and a cold wave getting mixed up in the atmosphere above Harlem, and the resultant hailstorm, whereof each crystal-clear stone had imbedded in its heart a prisoned locust.

"Just fawncy!" said his lordship. "I never knew such stohms weah common heah!"

Then before the startled eyes of the travelers a marvel grew. The visitor pocketed his monocle, shut his notebook with a bang, smoothed the look of vacuous imbecility out of his face, and became at once as much a business man as any of them. "Thank you, gentlemen," he said. "My friend Hart gave me to understand that you could tell me something about cold weather, and I have been most interested, but I have myself seen, in the Canadian dominions of Her Gracious Majesty, Queen Victoria—God bless her!—something surpassing anything you have told me. I was riding, one intensely cold night, on a Canadian Pacific railway train, between Caribou and Musquash stations, when all at once a tremendous crash occurred, and we found our train split into kindling wood and piled up all around the station of Musquash, which was very convenient, as we burned most of the wreck before morning in the station stove to keep from freezing."

"How did it happen?" asked Hart, while the other men in the group stole uneasy glances at each other.

"The engineer miscalculated the distance," said the visitor, his crisp utterance contrasting strangely with his assumed drawl a moment earlier. "It was thirty-five miles between stations. He was accustomed to stop before reaching Musquash, and back onto a side track to let a long freight pass. But on this night the cold shrunk the rails of the track so much, and they were so firmly plated together, that they drew the two towns a mile and a quarter nearer together, and the engineer, mistaking the distance, crashed into the waiting freight.

"Why, up there I knew a man killed by his best friend, in a fight over a dog," continued his lordship, with a glance at the stupefied face of Dr. Binninger. The visitor and Hart were apparently the only men in the room enjoying themselves.

"After he knew the facts, the survivor was simply crazed with grief at what he'd done. The dead man had owned a magnificent Newfoundland, and accused the other of stealing it. The accused man didn't think it worth while to mention, until after the fight, that he had rescued from the cold and taken home a tiny pup about seven inches long, of breed unknown to him. When he returned home he found in his shack the lost Newfoundland, and didn't at all understand the situation until I suggested to him that the Newfoundland and the tiny pup were the same animal, and that the extreme cold had shrunk the magnificent creature, about whom the quarrel occurred, to the mean dimensions of the little animal he had picked up in the snow."

There was silence for a moment, none of the members thinking of anything to say which would be adequate to the occasion. Finally Eckels changed the subject : "Dr. Binninger, what did you say to that new waiter the other night? The steward tells me that Henry ran upstairs in a perfect panic of terror, threw all of his clothes that he could get hold of, without too much difficulty, into a grip, and ran away, and has never been seen since. He didn't steal anything. He even left some of his own property, to wit: one large bone collar button, four razors, a concertina, a set of bones, and a plug hat. He seemed as if in terror of his life, and has vanished as completely as if the earth had swallowed him."

"I didn't say a thing to him," protested the virtuous doctor; "not a thing. The unfortunate youth merely stopped me at the door to ask some trifling, unimportant question. I myself noticed the extreme agitation of which you speak, but am entirely at a loss to account for it. However, if the members think that I am in any way responsible for the poor boy's disappearance, I will endeavor to look him up and report on the matter. A Southern gentleman of the old school, suh, never deserts a poor colored boy in misfortune."

"Think ought state," said Hart, in his queer, jerky way, "took privilege of Travelers' Club, introducing my friend as English lord. No lord at all. Just plain Canadian lumberman, Donald Fraser of Assiniboia. Gentlemen, again let me introduce my Canadian friend, Mr. Fraser."

The customary murmur of recognition went along the circle, and Mr. Fraser, who alone seemed to have the power of speech left him, bowed low, and said in his mellowest tones: "I am sure I am pleased to thank you all, gentlemen, and to express my satisfaction at having spent a most enjoyable evening."

CHAPTER VI.

MARCH GALES AND OTHERS.

"Wow!" said Jim Hart, stamping his feet as he banged the door of the Travelers' Club behind him. "Worst March gale ever I saw. Curious how it catches moving vans. On my way here, saw two of 'em, great big fellows, poking along street, kind o' careful like, when biff! came little harder gust than usual, caught hind wagon, keeled it right over first one. Then first one left behind; wind keeled it over other. Away they went up street, first one, then other, like boys a-playing leapfrog. Never saw any one in life so scared as drivers were. Both of 'em jumped out when felt old ark begin to hump up. There they stood on sidewalk hopping up 'n' down, 'n' yelling. Wasn't anything else they could do, of course, but—"

"Horses killed, I suppose?" said Dr. Binninger, as Hart struggled with his overcoat and rubbers.

"Nope. Never touched 'em. When wagons jumped, they snapped traces, 'n' each time wagon came down, it fell kind of slanting, rolling over other, 'n' hit ground far enough ahead miss horses.

73

After two or three narrow escapes, they scooted out
sidewise, ran away. Furniture must have been
little mixed up inside, though; last I saw of wagons,
' God Bless Our Home ' motto sticking up over door
of one of 'em, just natural as boarding-house.''

"Yes," said Parker Adams, " it is a pretty lively
storm for the city, but of course nothing to what
you see sometimes out in the open prairie, where
there's nothing to break the force of the wind.   I
remember being on a slow passenger train in Kansas,
at one time, when a gale came up behind, and began
to shove us along the track like mad.   In a single
minute, during which we passed by actual count 713
telegraph poles, ten rods apart, the axles heated and
swelled, and the wheels stuck fast, and began sliding
along the rails without turning.   Then the wheels
and rails alike were transformed in an instant to
glowing masses of red-hot metal by the tremendous
friction, and the cars would certainly have been set
on fire, adding new horrors to the situation, if the
engineer, with rare presence of mind, hadn't let
down his water scoop as we came to a trough.   The
water splashed sidewise upon the red-hot metal,
cooling it in an instant, and, as it cooled, welding
the wheels to the rail, as well as each rail end to end
with the next for several rods.   The running gear

stopped at once dead still, firm as the everlasting hills, but of course the car bodies, going at such a rate, slid right along off the trucks. It all happened in a minute, but Lord ! Wasn't I scared !"

"There must have been a calamity when the cars descended," exclaimed the ex-Kentuckian, Binninger, who was mixing some things in a big glass bowl.

"No ; not a bit of it ! You see, when the water from the scoop struck the red-hot rails and wheels, a tremendous mass of steam was generated. We were in a cutting at the time, and the steam blew us out of it like a cork out of a bottle, throwing us only a few feet above the natural level of the land. The wind was coming kind of quartering, a little from the right, and it just slid the car bodies off over to the left, and there we settled down as nice and easy as a rocking-chair. But I shudder to think what might have happened, but for that steam."

"You'd 'a' gone right against the left hand bank, bang ! as nearly as I can judge from your description," remarked Tom Fenton. "I never had any such narrow shave myself, but my hair did stand up once, I tell you, when I was in Texas. It was a breathless, hot afternoon, no wind, black clouds— well, you know what it's like, just before a norther —and I was riding range with three or four cow-

boys, fourteen miles from shelter. Then came the
wind, wooh ! And all at once the black clouds
turned white, and began chunking us with the darn-
edest, biggest hailstones I ever saw in my life. Most
of 'em were about the size of a football, but, once

in a while, a fellow as big as a rain barrel would
come along—two or three stones stuck together, as
near as I could tell afterward. Well, gentlemen,
this sounds rather thrilling, but as I'm a man of
truth and honor, the very size of the hailstones was
our salvation. If they'd been smaller, we couldn't

'a' dodged 'em anyhow. As it was, we jumped off
our broncos, which were struck dead in an instant,
and began looking up and dodging the lumps. They
were so big we could see 'em coming a hundred
feet up, and get onto their curves. Of course, too,
such large hailstones couldn't be very close together, and we were all slender young fellows and
active, and could get out of the way. But one of
the cowboys—poor fellow!—was a Harvard gradu-
ate, and as shortsighted as a second fiddle. Right
when we were hopping about, like hens on a hot
griddle, his glasses fell off, and he couldn't see a
thing; no sir, not a thing! It was horrible to hear
his one agonizing shriek ere he fell, and not to be
able to help him, but we couldn't look down long
enough to get his lamps for him, and—I don't like to
recall the scene. It didn't last long, but—well, to
cut a long story short, after the bombardment stopped, we set up a heap of hailstones to mark the
spot, and went away. We had to go 150 miles for
an undertaker, and when we got back, three days
later, the body lay frozen within its icy monument,
but about it all traces of the storm had passed away,
and the summer sun was smiling down as sweetly as
if death were impossible, and storms could be no
more."

" It must be a terrible shock to you, even now, to
recall such a scene, said Dr. Binninger, " may I ven-
ture to suggest, Colonel, that—" he waved a drip-
ping punch ladle in the direction of the glass bowl.
For a season brief but ecstatic, the members gathered
about, and the gentlest of aromas, a flavor, subtly
insidious, of mingled rye, lemon, sugar, and other
benign vegetable products, was wafted throughout
the room.

" Speaking of punch," said the big Kentuckian,
setting down his glass with a long sigh; " speaking
of punch—and I trust it may not seem the arro-
gance of self-esteem, if I remark parenthetically, that
this punch, that we have just enjoyed, is most excellent—I wonder if I ever told you of the time when
one of the largest cellars in Lexington was tempo-
rarily transformed by the direct visitation of the
Almighty into a perfect sea of unparalleled punch.
It happened that a thriving wholesale grocery, in that
enterprising town, was struck by lightning, which
completely fused the varied contents of the seven
stories. It also happened that the stock in trade
consisted, as is usually the case in our state, of arti-
cles imperatively demanded by the human constitution in these latitudes, such as whiskey, sugar,
oranges and lemons; and these ingredients were

also most naturally present in about the proportion,
in which they are in usual demand. The lightning
broached every cask, smashed every barrel, crushed
every lemon, and so tore the floors that their contents speedily sought the cellar, which was fortu-
nately equal to the responsibility. The amiable fluid
was there joined by a little water from the angry
clouds—though, for myself, I could never see that
good punch was improved by the addition of the less
noble fluid—and the comparatively insignificant
quantities of spices, fresh fruit, and other elements
which had been in the emporium, added to the mass
a certain piquancy much praised at the time by con-
noisseurs. Many people partook appreciatively of
the punch, and several of Louisville's talented clergymen—than whom the South has no more gifted, suh
—preached upon the incident as an eloquent refuta-
tion of the theories of certain fanatics, who hold
that the Author of the Universe looks, with dis-
pleasure, upon convivial enjoyment of the blessings
by Him vouchsafed."

"After all," Fenton broke in upon the silence
that followed the relation of this incident, "it
couldn't have been better punch than this."

"Suh," said the Kentuckian, with a low bow,
"punch past is never equal to punch present."

" The fortuitous mixture of the punch ingredients reminds me," said Parker Adams, " that I once assisted at a thunder-storm in Oklahoma. After it was over, the late householders were greatly surprised at finding, upon the supposed site of the extinct grocery, a large mass of excellent ice-cream in bulk, melting rapidly away under the sun, but still good at heart. The explanation was simple. The lightning-stroke which destroyed the roof of the building, and shattered every barrel and bottle in the place, fused and melted a dozen milk cans, releasing their contents. Directly over the cans, on the shelves, were a number of paper bags of sugar, a sack of flour, and seven bottles of vanilla extract, whose released contents fell into the mass. Before the milk had time to flow away, it was buried up in such hailstones as even Oklahoma never saw before, a fall of two feet occurring almost in an instant. The ice balls mingling with the contents of a dozen barrels of salt, which had been standing about the milk cans, produced such an intense cold that the mingled milk, vanilla, sugar, and flour were instantly solidified on the surface, and, in half an hour, became a firm mass to the core of excellent ice-cream."

" If it could but come more gently," sighed Dr.

Binninger, "the ice would no doubt have been highly appreciated in a warm climate like Oklahoma. Ice is even more scarce in Texas, where a remarkably successful plan of obtaining it has been invented, by a very dear old friend of mine, living near El Paso, named Barney Medary. Even in the hottest summer day Medary will partly fill a tin can with water, attach it to the tail of an enormous kite, and send it to the great height of three miles, where, in the rarefied atmosphere, it is promptly frozen, completely filling the can. After a sufficient interval, the kite is hauled in with a windlass, and the cake of ice removed. But the inventor is now perfecting an even more effective plan for manufacturing on a large scale. He sends up a huge kite, 150 feet long, by a steel wire cable. Under the kite is suspended a strong pulley, over which runs an endless chain, bearing at intervals hooks, on which the cans of water going up, are hung. As they just balance the cans of ice coming down, but little power is needed. As each hook passes the operator, he removes the can of ice, and hangs in its stead one of water. The cable travels but three miles an hour, and as the big kite is kept an altitude of from four to six miles, the cans remain in the cold strata of the air quite long enough for the water to become solidified. The inventor expects to be able to furnish a ten-cent lump of ice for three cents."

"I happen to have got wind," said Parker Adams, "of an enormous scheme for modifying the climate of the Eastern coast of North America in the summer time. I expect to make a fortune out of a few shares in the company. Some of the more prominent investors are John James Pastor, Cornelius Scott Vanderbilk, Drechsel, Dorgan & Co., and others whose names are a synonym for probity and vast resources. The syndicate has acquired a number of townships of prime, glacier-bearing land in the vicinity of Rejkajvik, and has contracted with shipbuilders, Clamp & Sons, of Philadelphia, for a fleet of powerful ocean tugs. The scheme is, in brief, to tow icebergs of manageable dimensions down from Rejkajvik, and deposit them at regular intervals all along the coast. Then, by means of powerful blast exhausts, worked by tidal power, the entire air supply of a dozen sea-coast states is to be passed over the cooling ice. For use on navigable rivers, like the Hudson and the Delaware, ice floes, drawing no more than a dozen feet of water, are to be employed.

"Ice grottoes in the bergs, after the manner of

those in Swiss glaciers, are to be inexpensively cut, en route, by the steam-jet process, so that when the bergs arrive off the coast they will be ready to receive visitors. Off Newport and Bar Harbor there will be first, second, and third-class grottoes, after the English fashion. The cost of all this might seem almost prohibitive, but the syndicate is also perfecting a patent by which a cheap and powerful engine can be put on board an iceberg, converting it for a time into a real automobile ice steamship, with its own propeller, capable of steaming four knots an hour, with hydrogen fuel extracted from the atmosphere by a new process. It is expected that by the summer of 1897 the climate of our Eastern seaboard, from Charleston to Mt. Desert, will be mild, balmy and equable."

"Hem! I am not sure that I could give such a scheme my unqualified indorsement," remarked Dr. Binninger, sagely shaking his head. "I question your legal powers in the premises. Many like hot weather, and will not thank you for destroying it. Besides, will not your process, by robbing the coast of its summer heat, make the winter's cold even more unbearable. Is it not bad enough as it is? Why, only a few weeks ago, a friend of mine, George Jenks, of Centralia, Kentucky, drew a quantity of boil-

ing water to take a bath. The tub was cold, and the
water cooled so rapidly on coming in contact with it,
that when Jenks jumped in, he was almost instantly
frozen solid in the tub, and perished miserably.
The wind, at the time, was blowing so hard that it
struck a copper kettle, and blew it inside out, so that
the legs were inside; and twisted the well so crooked
that you couldn't look down it without getting dizzy."

"I say, Eckels," said Parker Adams, suddenly,
"have you no information on this subject to add to
the common stock?"

All eyes turned toward a chair away at one end of
the semi-circle about the hearth, where, well in the
shadow, a lean figure was seen, chin on palms, and
elbows on knees.

"Somehow I don't feel in a scientific vein this
evening," Eckels responded, moodily.

"You look like a pretty fair liar, too," said
Adams.

"Appearances are deceptive. The honest waggle
of a meek dog's tail doesn't denote absence of
teeth. A horseshoe isn't lucky when the equine
owner propels it against your abdomen with his fairy
hoof. In short, I'll be excused, please."

"What is it, old man?" Harry Porter whispered
with ready sympathy.

"Nothing," said Eckels, shortly.

"Nothing! Great heavens! is it so bad as that?" exclaimed Porter, unconsciously quoting Eckels' own words on a previous occasion.

"D—n!" growled Eckels, jumping to his feet in a rage, "if I can't come here and sit quietly without being pestered by old fools with requests for stories, and by young ones with stale maxims, I'll get out, that's all."

A minute or more elapsed after Eckels had banged the door behind his retreating form; then Parker Adams emitted a long, low whistle and casually remarked: "Quite stormy, isn't it?"

They agreed that it was.

CHAPTER VII.

DAUGHTERS OF EVE.

Miss COPELAND and Miss Ann Copeland were as much alike as two women ever are, and that is not at all. Miss Copeland had hair which her admirers called "sunny" and her detractors "white-horse reddish." It was beautiful, it was abundant, fine, long and smooth; yet it was not her only charm; for with it went a plump though petite figure, a perfect complexion and big wondering eyes.

Miss Ann Copeland was of that moderate brunette type which is almost distinctly American. She was taller and more athletic than her sister, her complexion had more of the olive tint, and her big, brown eyes seemed in some lights almost black, like her hair, which in the sunlight was seen to be of a warmer hue.

The Misses Copeland, being unfortunately fairly well-to-do, had "afternoons" for the want of better employment, when, with the aid of other like circumstanced young women, they so managed to pour tea, squeeze lemons, handle tiny wafers, and otherwise play hostess, that each movement was a poem, and each pose a picture. It was like a child's game of "play house." A healthy tramp could have eaten and drunk everything in the room, and gone away hungry; but that wouldn't have been so poetic.

Eckels was shameless. He permitted Harry Porter to drop in at the Copeland house under his own guidance. It was not in accordance with his plans for Harry, but his own confidence, in his power to please Miss Copeland, was somehow shaken, and he wanted to see her again.

When the two were ushered into the parlor, where already the shades were drawn, and the candles lighted, though the March sun had not set, they were amazed at hearing, instead of the susurrus of half a dozen women's voices talking at once, certain big, booming tones they well knew. Dr. Binninger was seated, teacup in hand, holding forth to eight or ten pretty girls. There was no other male thing in the room save a tall boy of seventeen, who would have been more comfortable if he had known what to do with his hands.

When Eckels and Harry had greeted Mrs. Copeland, an admirably quiet and self-effacing American mother; when the daughters of the house had met them with dignified reserve, and the other girls, with more effusion, Essie Terburg, a tiny person with a Roman nose and a retroussé disposition, cried

in a shrill treble : "Oh, Mr. Eckels, Dr. Binninger
has been telling us such a funny story about the
English lord who fooled you so at the Travelers'
Club."

Eckels was angry to think that Binninger had
been putting him at a disadvantage before the
ladies, yet in his retort he was unable to think of
anything more brilliant than the familiar " you're
another."

" I'm sure," he said, "that Dr. Binninger was as
active in giving Mr. Fraser information, and as
much astounded when the tables were turned, as any
of us."

"Ha ! Ha ! Ha !" roared the big doctor ; "so I
was, my boy—so I was ! But it was a study to look
at your face. I wish the ladies could have seen you
when Fraser told that story about the shrinking dog.
Though to be sure," Dr. Binninger went on, becoming more grave, "it was no more wonderful than
many things which have come within my own
observation. Hem ! It was once my fortune, for
instance, to see a most terrific battle between a
rattlesnake and a puff adder. The rattler was so
strong that he quickly swallowed the other, but
then, so far as appeared to the interested spectator,
the puff adder began puffing. At any rate a great

balloon promptly appeared to be distending the rattler, and his skin seemed on the point of bursting.
Distracted by his frightful torture, he rushed into the
river, when the puff floated, naturally depressing his
head, until he was drowned. Of course the puff
adder perished with him, but had at least the satisfaction of selling its life as dearly as possible."

"I don't know why it is," said Nina Markham,
a tall and darkly beautiful girl, with the excessively
feminine air that big women are apt to effect ; "if
I were to see a snake, I should faint right away ; I
just know I should. And yet I do like to hear
about them," she added, reflectively.

"Possibly you would be interested, then, in the
dramatic suicide of a snake which I once witnessed
in Limerick, Ireland," continued Dr. Binninger.
"This was a large black-snake which, having been
worsted in a fierce battle with another of slightly
different species, trailed away in deep dejection.
Finally as it dragged itself along, closely watched
by Patrick McClaughry, a manufacturer of she-
been, and by myself, resolution seemed to fire
the soul of the defeated reptile. Grasping firmly
with its mouth a small stone, the snake climbed a
tree, and presently hung by its tail from a horizon-
tal limb. Next it began whirling about the limb

with frightful rapidity. Longer and longer its body
stretched under the centrifugal strain, until, with a
last despairing effort, the snake's body broke in
halves, the weighted head and neck flying to a considerable distance, while the tail remained clinging
to the limb of the tree. Then only were we able to
see the full purpose of the act. For just as a boy
throws a stone from a sling, the snake, in its dying
moment, released the stone it held in its mouth.
The missile, hurling through the air, struck with
deadly force the rival snake, which had followed to
gloat over the suffering of its victim, and crushed it
dead.

"Probably the most peculiar varieties of American reptilia are the hoop-snake, the glass snake, and
the sidewinder," continued the doctor. "In a
museum which I used once to visit, a glass snake
and rattler were confined together in a cage. The
rattler used to bite the other, but the glass snake
would promptly unjoint itself, and reunite without
the bitten section. By and by, when the missing
piece had recovered from the bite, the glass snake
unjointed and took it in again. But one day the
rattler, after biting its room-mate, formed a new
resolution. Before the glass snake could resume
possession of its missing sections, the rattler

swallowed it, and the glass snake was forced to do without a portion of its length. Again and again, this operation was repeated, until there was nothing left of the glass snake but its oddly joined head and tail. Here the two enemies were separated. The foreshortened snake, however, soon died, its constitution not being equal to the strain."

"Do you believe in the sea-serpent, Dr. Binninger?" asked Miss Copeland, who had scarcely addressed a word to Eckels since his arrival.

" Do I believe? My dear young lady, I once saw one, off the Solomon Islands, which had a huge open circular mouth like an enormous life buoy, with broad white lips and a dark interior big enough to admit a ship's boat. The beast was twelve or fifteen feet wide, and had several pairs of wing-like flappers. The fins were forty feet long, and the whole aspect was so generally horrid that extra grog was served, and there was less profanity aboard for twenty-six hours.

"Then there was a sea-serpent of undoubted authenticity last summer, off Block Island,—only a little one about twenty feet long,—and one in the Indian Territory, which had a head like a cow, ears like a mule, and a tail like an alligator. It was seventy-three feet long."

"Pshaw!" said Mrs. Copeland, "there isn't water enough in the whole Indian territory to float a snake so long as that."

"Quite true," said Eckels, "but a sea-serpent a

mile long can be floated in a pint of Kansas whiskey smuggled over the border in a boot-leg."

"I choose to assume," said Dr. Binninger, rather stiffly, "that Mr. Eckels has forgotten that I made the statement to which he apparently takes exception, upon my own authority. To resume the subject, the adaptability of the animal kingdom to the acquired habits of man, and the imitative instinct of the higher animal types, have often been noted by scientific observers, like Agassiz, Buffon, and myself. An interesting case is reported from Pike County, Pennsylvania. In old Pike, as in many other places, bicycles have become quite numerous, and have evidently inspired the mute onlookers of the woods and fields to emulation. The other day a Jericho amateur secured a photograph of a living bicycle. Its wheels were a couple of hoop-snakes, which, before forming a circle, after their usual manner, by putting their tails in their mouths, lined themselves up so that a gray squirrel could scamper upon their circumference as he would in a whirling-cage. With his forefeet upon one animated wheel, and his hind feet upon another, the squirrel could maintain its level, and carry upon its own back a seated chipmunk. This queer combination is capable of a speed of one mile in fifty-seven seconds—only six seconds slower than the best time of an unimpeded hoop-snake."

Some of the young ladies, quickly tiring of the subject, had begun a *sotto voce* conversation about art embroidery, and the quick ear of the doctor had noted the fact. Not wishing to lose an auditor, he turned to the dissidents with his blankest smile :

" I suppose you have heard," he began, " of well-authenticated cases, where a needle, entering a person's body, has ' traveled,' or ' worked its way,' from point to point, as the ordinary, unscientific person usually describes the process. These cases have usually been those of persons with very soft flesh. A girl in Mobile, whom I knew very well, carries this flabbiness of muscular structure to an extreme. Needles thrust into her flesh seem to cause her no pain, and little annoyance. She is accustomed, wherever she goes, to carry several needles, threaded with silk of different colors, thrust into various portions of her anatomy. To get one of them, she has only to hold a strong magnet upon the opposite side of her person, and the needle required is attracted right through her body."

Others dropped in, as the afternoon drew toward evening. Langdon came, with his wife. Tom Fenton "ran round," on his way home. There were two or three more awkward youth, "deyv'lish fellows, don't ye know," but subdued in company. And there were more girls.

Presently Harry Porter found himself talking to a washed-out blonde in baby-blue, named Curtis, while Ann Copeland sat within earshot. This was the situation Eckels had outlined when he said, " Talk at her, not to her."

" Ah, your sex is so selfish," Miss Curtis was saying.

"Yet for women men have died," said Harry, mentally resolved to talk his best.

"Oh, no; for indigestion, maybe." (Giggle.)

"I solemnly assure you, Miss Curtis," said Harry, gravely, "that there's a woman in Atwood, Illinois, for whom many men have died. She is an undertaker."

"Oh, you cynic!" and Miss Curtis playfully tapped Harry with her fan.

"No man can be a cynic in your presence," said he, warmly, bending toward her ever so little. "And no true woman can be a cynic at all, except as a pose. I never saw a cynical woman. This is why even the most acute of your sex so readily listen to the unworthy members of mine. And what man is worthy of woman? We are such feeble folk. Miss Curtis, a star once looked down on a city. It saw little creatures running about. They were born, died, fought, hated, kissed, loved, laughed, wept. But whatever else they did, they ran about."

"Ah, but some men are not feeble," said the little blonde, with a meaning look at Harry's athletic shoulders.

"When a man brags of his power, ask him to make a blade of grass. Ask him why we use a spider's web, and not any thread that human ingenuity can spin, to mark off the objectives of our telescopes. Ah, the folly of man! When he would recreate himself, he does not consider how he can make himself different. He only thinks of going to see some new thing."

"Really, Mr. Porter," said Miss Curtis, "you are so critical of men, what must you think of women?"

"I? I am like a violin, that when pressed by the hands of beauty, can emit only hoarse, unmusical sounds. I am always at my worst when, with a beautiful woman, I would be at my best."

"But is a woman always at her best?"

"Possibly; probably; I do not know. At least it seems so to me when I am with her. I—"

"Miss Curtis," said Ann Copeland, hurriedly, her clear cheeks aflame; "there's a nice boy over there, who is having a horrid time. I wish you'd let me bring him over for an introduction."

Eckels, meanwhile, had made a last desperate attempt to regain control of the situation, in another quarter of the room, by telling Miss Copeland and Nina Markham a fairy tale about the shrinking propensities of new flannel garments, a subject somehow suggested by Dr. Binninger's needle story. "Gus Tooper, of Alameda," he said, "had put on a very heavy new flannel undershirt, which was rather tight, and began work, chopping wood. Profuse perspiration ensued, and Tooper presently complained to his companion, of a strange, oppressed sensation. Then he fell, in a dying condition. The inquest developed the fact that he had expired from the pressure of the shrinking undershirt expelling the air from his lungs.

When Eckels had lamely finished his bungling tale, Miss Copeland looked at him rather queerly for a moment, and then changed the subject, deftly conveying the idea that the narrative bordered on indelicacy. Nor would she thereafter notice his presence.

So, with much chattering, the callers gradually drifted away, and the sisters were left together.

"What a splendid, intellectual fellow Harry Porter is!" said the elder, presently. "I don't see why—"

"That's just what May Curtis said! The little goose! I know she didn't understand a word he was saying!"

"Well, you need n't be so spiteful about it. You could have had him, if you wanted him. And, of course, he is brilliant. He was in Princeton."

"I don't want him! And as for Dr. Binninger, I despise him!"

"Still," said Miss Copeland, "a man's a man; and they're none too plenty at 'afternoons'!"

"No wonder. Men have something to do! I wish I were a man! If I were one I wouldn't let any little shrimp of a woman snub me, as you did John Eckels to-day. Such a splendid fellow as he is, too!"

"I don't think you're a good judge of men. You threw over Harry Porter."

"I know I—dud-dud-did! Ah, hu! hu! hu! hu! hu! hu! I wish I were dead!"

But, surely, no such conversation ever took place; ever could have taken place.

This is a Book of Lies.

CHAPTER VIII.

LOVE HATH MURDERED SLEEP.

THE birds had all day perversely preferred whole skins and freedom to the honor of reposing in their game bags, and the detachment of travelers, who had been tramping the Jersey marshes, were revenging themselves over their steaming glasses by tales of achievement in past campaigns.

"The most remarkable bag recorded in the annals of my memory," began Dr. Binninger, " was obtained by unsportsmanlike methods, which were promptly and properly punished. It was in 1875, when water fowl were very abundant. Not content with slaughtering wild ducks by the usual methods, Harry Jones, of Currituck, procured a Gatling gun and loaded the cartridges with bird shot. Concealing himself in a blind, he waited until a large flock approached him at great speed with a favoring wind. When they had nearly reached him he opened fire. The destruction was terrible, but such was the impetus of the birds and the force of the wind that almost the entire flock, which he had slaughtered, flopped dying upon his hiding-place, and, beneath their mangled bodies, Jones perished miserably by retributive suffocation."

"I know a fellow named Louis Schrempp, St. Louis, had great luck shooting squirrels one day '79," said Jim Hart. "Went out with gun, Louis did, myself and two dogs. Presently dogs began spinning 'round 'n' 'round; kept it up till fell down exhausted. You see, so many squirrel-tracks, couldn't tell which to follow. Then Schrempp and I looked up. Trees full of squirrels, I was utterly useless. Got dizzy watching dogs whirl about; but Schrempp shot away all his ammunition, then began picking shot out of dead squirrels to begin again."

"But," said the doctor, "he couldn't—"

"No, he didn't. Too slow altogether; weather pretty hot. Gave it up. Picked up 157 squirrels, brought me and the dogs to our senses—"

"Yes?" doubtfully; by all.

"—To our senses; then went home."

"An old Indian, who lives in Carson, Pennsylvania, once told me that a rattlesnake will always range himself in line with a stick or gun pointed at him," said Eckels. "This peculiarity makes it easy for even a blind man to shoot one. The Indian took me out with him, and selected a nice big snake. Whenever

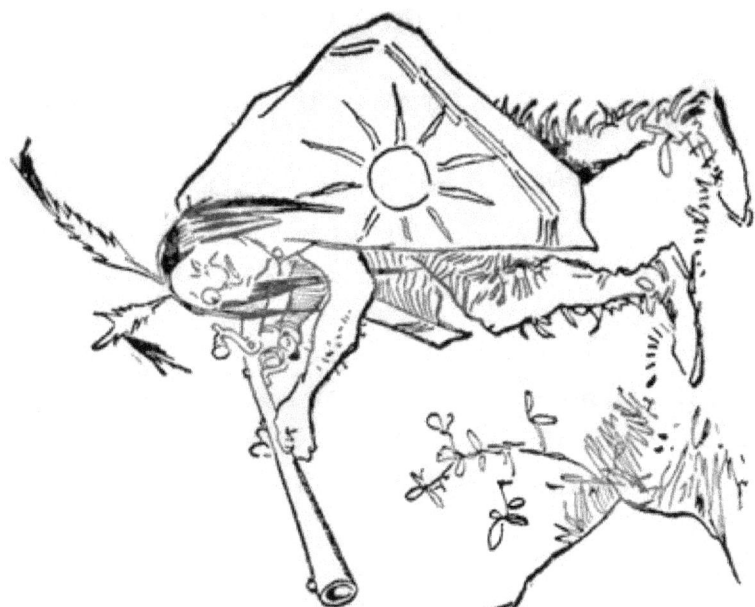

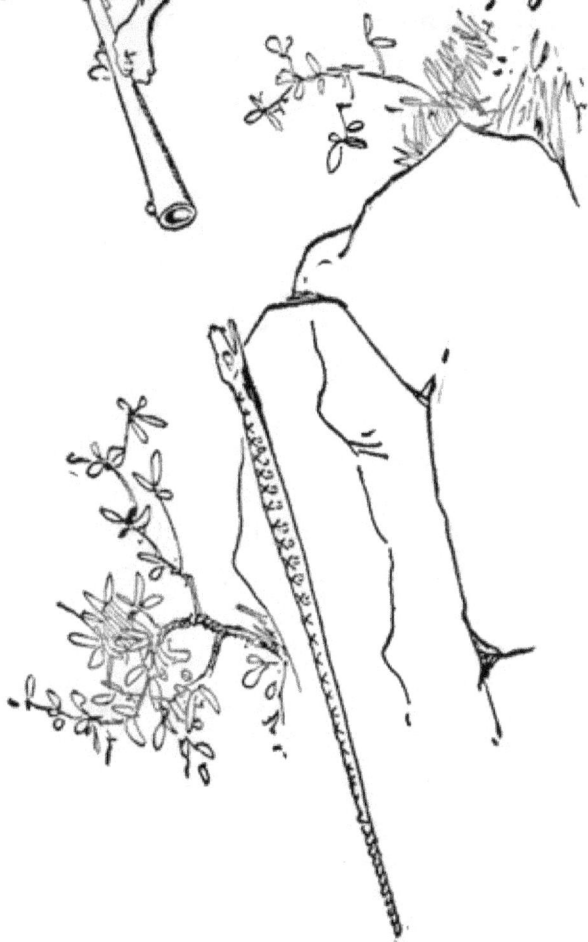

the Indian moved his gun, the snake would get
in line.   Finally Lo fired, and his bullet went in at
the snake's mouth and passed through the entire
length of his body.   Indians have no imagination.
I wouldn't have wanted to do what he did with the
body.   In the dry air of the mountains, flesh doesn't
decay, but dries up as hard as a brick.   The Indian
coiled the hollow snake up, and when it had set in
that shape quite hard, he used it for the ' worm ' of
a moonshine distillery he ran, up in the hills.   Ugh !
No wonder the fellows who drank that whiskey saw
snakes ! "

After the libation suggested by the thought had
been carefully swallowed, Tom Fenton began a
reminiscence of travel.

"The African steamer *Winnebah*, on which I was
a passenger from Liberia to Oporto," he said, "had
a singular passage, in which sudden alterations in
the weather, from very hot to very cold, preyed
upon the superstitious fears of the sailors.   Off the
Morocco coast, the ship sailed for sixty miles
through a mass of locusts, covering the water to a
depth of several inches.   Many of the insects were
seven inches long.   They had been blown out to
sea by a strong, hot wind, and a sudden cold wave
had killed them all."

" The stench must have been frightful in the hot sun of those latitudes," said Dr. Binninger.

"Not at all. The cold wind which killed the insects evidently came from Labrador. It was, at any rate, heavily laden with powdered lime rock, which lay in a thick dust upon the bodies and acted as a disinfectant.

" The propagation of insect plagues is a fascinating study," Fenton continued, tossing back his theatrical locks, slightly tinged with gray, from his pale, fine face. "In the town of Quantuck, New Jersey, which lies in a low, hot nook, surrounded by swampy land, the mosquitoes were so thick in the season of '83 that, when the breeze was gentle, they formed a thick, black cloud over the town. On several occasions this was so noticeable that the hens went to roost at noon, under the impression that it was already nightfall, and without performing their daily task of egg-laying. As the poultry business is a leading one in the town, the fanciers suffered for a time considerable financial loss, until the device was hit upon of sending up small dynamite cartridges among the thickest swarms of mosquitoes, by means of a kite, flown by a wire, which, at the right moment, conveyed a current of electricity to discharge the dynamite. After a few discharges the air would be so clear that the hens could resume operations, and the gory remains of the dead mosquitoes, falling to the ground, were plowed in as fertilizer."

" The modification of the lower animate orders by natural causes is another interesting study," said Parker Adams, wearily stretching his short legs to the blaze. "It is well known that heat expands and cold contracts ; but to an even greater extent rarefied air causes expansion of soft bodies by relieving them of a portion of the pressure of the air. This fact, well understood by balloonists, is illustrated by the enormous size attained by the mosquitoes on the top of Mount Orizaba. So large have these become, after centuries of living in cool and rarefied air, that it is a common amusement for rich and sporty Mexicans to catch and train two of them, and pit them together in a cage like wild animals. The ensuing combats are described as equaling, in intensity of terror, the fabled tiger matches of Eastern potentates, although of course the combatants are scarcely larger than good-sized squirrels."

" The intelligence of the so-called lower animals frequently puts to shame man's assumption that he is the only reasoning section of creation," said

Dr. Binninger. " Robert Hinckley, of Peoria, has a dog which has an unconquerable aversion to getting wet. Recently, desiring to cross the river to save a detour of several miles, this intelligent animal set to work to build a raft. Dragging stick after stick to the water's edge, he laid them side by side, afterwards crossing them with others until he had built a raft, frail, indeed, but amply able to sustain him. Then he pushed the raft into the water and jumped on. Paddling with fore and hind paws, he soon propelled himself across. Nor did the animal's foresight end even here, for, carefully towing the raft to a little bay, he secured it, until his return, by laying one end of a fence rail upon it.

" Quite as cunning, was a cat I once owned, which habitually provided herself with a meat breakfast by littering bread crumbs and burying herself in the snow until the snowbirds came to peck at them, when she leaped forth and smote them with triumphant paw. This instance of sagacity is quite surpassed by a cow in Oxford, Mississippi, which saved herself from freezing to death, during one of the frequent blizzards in that state, by swallowing a lump of freshly-burned lime and industriously chewing snow. The lime, in the process of slaking, kept the cow warm, while all the rest of the herd perished. A severe fit of indigestion was the only untoward result of the stomach-load of whitewash."

" I wonder how you happened to miss that shot to-day, Fenton," said Parker Adams, rather irrelevantly.

" In Fannin County, Georgia, there still survive large numbers of wild turkeys," said Fenton, hastily, affecting not to have heard the remark ; " Milton Ganthony recently discovered an enormous flock of these in the Splay woods, and, having no gun, was at a loss how to improve the occasion. Finally he happened to recall that he had a red bandanna handkerchief, which he tied to a twig and then withdrew, uttering a turkey call. When the splendid birds came and saw the handkerchief, they began fighting with each other fiercely about the meteor flag. When the war had gone far enough, which wasn't until Ganthony thought there were as many dead turkeys as he could dispose of, he shooed off the combatants. Besides a good wagon load of turkey meat, already plucked, he got loose feathers enough to make 573 turkey-tail fans."

"Ganthony was more of a hunter than you are, at any rate," said Adams, returning to the attack. " For my part, I find sport at the present day too tame. I'd like to have lived in the old times when animals were sure enough big ones. You know, of

course, that Montana was formerly inhabited by a
race of prehistoric buffaloes. Ephraim Rogers, a
well-known Butte trapper, owes to this fact his life.
While out of reach of his rifle, Eph was chased
by a grizzly. It was on a bleak and desolate plain
and no tree was within miles. Suddenly Eph espied
a hollow buffalo horn of enormous size, into the butt
end of which he dived in desperation. On came the
bear after him. Eph is a man of slender build,
while the bear's head and shoulders were very
massive, so that while the former easily crawled
out of the little end of the horn, it was carried
off by the blinded and infuriated bear, who could
not dislodge it from his head. That night a
mining-camp thirty-seven miles from the scene of
the adventure was put into a panic by the apparition
of an animal of curious appearance which, when
shot, proved to be no other than Roger's late
antagonist, still firmly fixed in its extinguisher
of buffalo bone, and unable to see where it was
going."

" By the way," said Dr. Binninger, " speaking of
buffaloes reminds me of the proposition to replace
the buffalo upon the Western plains by the hardy,
edible, domesticable, acclimatable and philoprogeni-
tive kangaroo. Old travelers in Australia can

recall numerous instances of the cunning of these
marsupials. In the Karaboo diggings, in the summer
of 1887, a gathering of kangaroos came to the bank
of the Waroo River, and began figuring how to get
across. Presently one of the smallest and lightest
went back from the bank a little way, and several
big fellows ranged themselves in line. Down came
the little one, leap after leap, and, as he reached the
line, each of the others in turn gave him a tremendous kick until, with the last one's final boost, his
momentum became so great that he landed dry shod
upon the further bank. Here he shoved a farmer's
boat into the water, and after that all was easy,
the entire party crossing the river in about seven
trips.

"Those who know the habits of the kangaroos
regard them as man's chief friend. When lost
upon the arid Australian plain, miles from visible
water, the prospector or herdsman always hails
with delight the appearance of a kangaroo well.

"The kangaroos, when they wish to dig a well,
gather in numbers and mark out a circle of about
three feet in diameter. Then each in turn runs and
jumps into the circle landing stiff-legged—ker-
chug!—upon his hind feet. In this way the hole
is sunk lower and lower, precisely as a drill sinks

into a rock with each successive blow. The loose
dirt is brought out, from time to time, in the pouches
of the diggers. A well has seldom to be sunk more
than eighteen feet to strike water, and at that depth
it is an easy thing for a kangaroo to jump in and
out again, though, in the case of a very deep well,
there is sometimes a platform half way.

" I never suffered from thirst in my life," Dr. Binninger went on ; " thirst for water, I mean. I do
not know what the sensation is. But a hunter may
face perils as deadly in hunting squirrels in the
Maine woods as in the heart of Africa. Once, under
conditions such as I have described, I reached
up my gun barrel to knock a lump of spruce gum
from a tree trunk. The gum lodged in the muzzle
of the gun, and thoughtlessly, though the day was
very cold, I attempted to pick the gum off with my
teeth. Instantly my wet lips and tongue froze to
the barrel, causing me the most excruciating agony.
I was far from home, and it was difficult to walk in
the constrained position rendered necessary by the
gun. Finally, as I stumbled through the bushes,
the hammer of the gun caught and it exploded.
My first thought was : ' Well, my brains are blown
out and I must be dead.' But I was not only alive
but sound, and released from the gun. The severe

cold had so contracted the barrel that the bullet, heated and expanded by the burning powder, could travel only about eighteen inches toward the muzzle, where it stuck fast. The breech of the gun simply blew out. The heat generated by the friction and explosion of the bullet a moment later warmed the gun-barrel just enough to release my frozen lips from the kiss of death."

"Extreme cold weather brings many dangers," said Fenton. "I was once in Kansas at a time when a crack in the earth was opened by the combined action of extreme heat on the inside and extreme cold on the outside. This fissure, which was situated near Union Star, was only about five inches wide, but emitted a smell like burning wool at least a yard wide. The people of the neighborhood tried at first to fill the fissure by throwing in various substances, but desisted upon hearing a faintly echoing voice ascend one day, freighted with the query : 'What the —— are you doing up there ! I'm from Kansas myself. Stop throwing them stones !'"

Sleep comes soon to eyelids weighted with fatigue. One by one the men about the fire nodded, and slipped away to bed, and presently Eckels and Harry Porter found themselves alone.

"You made quite a hit at the reception, Porter," said Eckels, presently. "That Curtis girl was visibly impressed by your wisdom. How in thunder did you manage it?"

"Simply by following your admirable advice. I thought up some brilliant sayings myself, and copied some more out of a newspaper and memorized them all. Then the other day, when I saw that Ann was listening, I just got them off one after another, whether they fitted into the conversation or not. And, by George, they didn't know the difference!"

"I should say not!" said Eckels, ironically; "The blue girl halfway fell in love with you then and there, and Ann came to the conclusion that you're a deep and dangerous person in need of reformation. Of course that settled her. You're all right if you'll only be patient. But did you see the Prince of Wales frost I got?"

"Don't be discouraged, Jack; follow your own advice. And now I think it's high time you told me what to do next. Isn't the field right for a direct attack?"

"No, not yet; but I will tell you."

And the fire burned lower yet before the two men sought their belated beds.

CHAPTER IX.

A WANDERER FROM THE WRATH.

A NUMBER of men from the Travelers' Club had stopped in the barber's shop on Fulton street, one day in early Spring, and the barber made the discovery that Parker Adams was becoming bald.

" I have here, " he said " a tonic that—"

" Tonics are a crude substitute for the resources which science is now able to bring to the relief of bald men," said Dr. Binninger.

" Wear a wig ? " asked the barber.

" I do not, sir; " and Dr. Binninger deftly adjusted it under guise of scratching his head ; " nor do I now refer to wigs at all. I have read, however, an American item to the effect that both wigs and tonics are quite out of fashion in France, even among the entirely bald. Holes are made in the head with a gimlet, and each hair inserted separately and soldered in its place. The process is said to be extremely soothing, and the result is charming, the metallic hue of the solder gleaming gaudily through the surrounding stubble."

" How know its American ? " demanded Jim Hart.

" Because I saw it in the *London Million*," said

the big doctor, shortly. "I am reminded by my surroundings that a unique occupation for lazy men has been invented by a Denver barber, who has a wonderful patent hair restorer, which I myself once used, and to whose aid I owe the remarkable preservation of my natural covering. He hires a number of ex-clergymen, who have left the East on account of chronic ministerial sore throat, to sit in armchairs all day long and grow heads of hair. The tonic is applied three times a day, and the clergymen are shorn once a weak. The hair is made up on the spot by expert wig-makers, and the refuse is used for cushion stuffing. The wages paid range from $9 to $16 a week, and found, and afford the rarest opportunity to combine recuperative leisure with remunerative industry, which it has been my good fortune to observe."

Here a slight commotion was observed in the rear of the room, where a colored bootblack had established his stand. The operator thereat had knelt on his own footstool and, with his face buried in his hands, was quivering violently.

"That boy's back looks familiar," said Dr. Binninger, reflectively; "looks like a likely boy I used to know, Bill Coons of Canton, Arkansas. Coons and a man named Bargelt were taking a short cut through the woods recently, hauling on a stoneboat a windlass they were intending to use in digging a well on the Milt Jones place. Passing a big, hollow tree, Bargelt saw a 'possum tail in the opening, and tried to pull the beast out, but met with unexpected resistance. It is well-known by old hunters that a number of opossums treed under such circumstances will cling closely to each other to defeat the efforts of the hunter.

" 'Set up the win'lass while I hol' on !' shouted Bargelt.

" In about ten minutes Coons had the windlass stoutly pegged out, and the well-rope was noosed round the hind quarters of the only 'possum in sight. But, strain as they might at the windlass crank, the two men could not make the 'possums budge.

" Now, thoroughly excited, they tackled the old horse to the windlass crank, and by uniting their efforts with his, they had presently hauled out of the tree a long rope of 'possums, each of which clung to the next one by winding his tail 'round his neck. After their manner, the 'possums feigned death until the last one had been killed with clubs, when they were loaded on the stoneboat for transportation home. There were 137 of them in all, yielding about 1,000 pounds of meat, which sold readily for seven and one-half cents a pound, net. The skins were worth twenty-three cents apiece."

The boot-black now shook so violently that the chair rattled.

"Sick, Henry?" asked a journeyman barber, with pompadour hair and a horseshoe scarfpin.

There was no answer.

"Ah, t'ell wi' the coon!" judicially observed the head barber. "Next!"

"The Southwest is indeed a land of surprises," said Dr. Binninger. "For instance, the skill of the vaqueros is marvelous. Juan Fernandez, of Santa Fé, was crossing the prairie recently, when he was met by a ferocious bull. At a critical moment his horse thrust its foot into a prairie dog hole and fell, leaving Juan on foot. Just as the bull was thundering down upon him in the charge, an opportune whirlwind passed near the scene, and was instantly lassoed by Fernandez. It whirled him away in a cloud of dust and dried grass, and, in a moment, he was far out of reach of the bull and headed nearly straight for home. Here, boy!" the doctor continued, walking back to the kneeling figure at the boot-black's chair, "Wake up, there! I want a shine!"

With a howl of terror, the boot-black darted toward the open door, his face livid, his limbs quaking.

"Here, George! Henry! Charles! Abraham Lincoln!" shouted the doctor. "There, he's gone. Did you see who that was? By the immortal George Washington, that was the very boy who ran away from the club! Now, what can be the matter this time? Why such precipitate haste to escape?"

"Aw, don't mind 'im," remarked the head barber; "'e's nutty. 'E's full o' dope. 'E's bughouse, sure. 'E ain't been a thing but scared of 'is shadder ever sin' 'e come 'ere. James, chase out 'n' git a boy to shine up th' gent. It's good tonic, all right, all right."

"Most curious how that boy acted," mused Dr.

Binninger ; "but speaking of baldness, did you ever reflect upon its cause and cure, and upon the erroneous notions current concerning it ? Some say it is due to the excessive eating of meat, but its more likely to result from worrying about how to pay for the meat. It cannot be due to wearing hats, for the hatless Romans and Greeks were frequently bald ; nor to a wicked and ill-spent life, for the bald theological student and fierce and hairy train robber disprove any such theory. Painters wear long hair, yet frequently represent the saints and apostles as bald. There are no bald heads in lunatic asylums. The whirring of wheels in the brain seems favorable to productiveness of hair on the scalp. I wish I knew to what cause to attribute the erratic behavior of that boy."

"Aw' 'e's dotty. 'E sees 'em again," said the head barber. "Is this an open game? Kin I come in ?"

"If you stand pat," said Hart.

"Say," said the barber, continuing his deft strokes as he spoke. "This bald business is queer, all right, all right. I'm a farmer, that's right. Come from Island Falls, Maine. The' was a big tannery there ; used up 600 hides a day. Say, the water in the tan pit was a hair tonic. That's right. Sprout

the hair on a billiard ball in three applications. Say,
the baldies use' to come a hundred miles to souse
the tan pickle on their cocos. Then they stopped
all at once ; found out the new hair was growing in
red, brindle, black and white, any old color. Say,
it was jes' like the cow's hair been soaked off in the
pit. Say, that's about when I lit out. Place too
big to hold me. Say, good old New York ain't in
it wit' no Island Falls. I guess nit ! "

"Your custom of expressing a proposition by
affirming its negative makes your narrative difficult
to follow," said Dr. Binninger, "but it seems a most
worthy and interesting one."

"The mention of Maine reminds me," said Tom
Fenton, "of a cat named Fanny, owned by a lady
in Thomaston in that state. This cat had kittens,
as cats named Fanny frequently do. Hearing her
mistress remark that the kittens must all be drowned,
Fanny removed them one day to a safe hiding-place.
After a day or two she brought them back to the
mistress with an air of triumph. It was then seen
that she had, with teeth and claws, torn to pieces an
old canvas and cork life-preserver, and had fastened
a piece of cork around each kitten's neck. Moved
to pity by the sight, the mistress said : 'Fannie, not
one of your kittens shall ever be hurt.' At this the

intelligent mother took off the bits of cork and went
to sleep in perfect confidence.

"The intelligence of animals is sometimes wonderful. There is a girl in Murfreesboro, Tennessee,
whom I know very well, as she is engaged to a
friend, who has a musical cat, trained by patient
practice to sing the popular songs of the day to the
lively accompaniment of the banjo. 'Ben Bolt' is
Kitty's favorite, but it is as good as a show to see
her march out of the room with a distended tail and
air of offended dignity when the chords of 'Say
Au revoir, but not Goodby,' begin to smite the air."

"If animals cannot reason," said Dr. Binninger,
"their inborn instincts frequently seem to differ
from reason in nothing save that they are even
more shrewd. A Manx cat of the gentler sex,
residing in Oshkosh, was recently confronted by the
necessity of saving four kittens from a spring flood
which menaced their home. There was no time to
take them, one by one, to a place of safety. But
not for an instant did the intelligent animal hesitate.
Leaping high from the floor, she knocked down
from its hook a cap worn at school by the boy of
the household, and this she placed bottom up on
the ground. Then, placing the kittens in it, one by
one, she seized the visor in her teeth and hurried

away to a place of safety, dragging after her her extemporized baby carriage.

" The mention of cats naturally reminds me of their hereditary foes, mice and rats, which are animals quite as shrewd.   It is a fact well known to scientists that even the lower forms of nature follow to some extent the peculiarities of the men with whom they are thrown.   Naturally in Connecticut, where men are wise beyond the ways of their kind elsewhere, lesser beasts have also wisdom of a high order.   A mouse — this happened in Norwich — which, I am told, is pronounced by the inhabitants so that it rhymes with ' porridge '—finding himself caught in one of those traps from which egress is discouraged by a *chevaux de frise* of pointed wires, spied, on the floor, near the trap, a piece of stout string and a toothpick.   Painfully he succeeded in clawing them into the cage, and, looping the string about one of the wires, he attached the other end snugly to the wall at one side.   Then, applying the toothpick in the manner of the stick in a tourniquet, he twisted the string until the fibre torsion bent the pointed wire back out of the way.   Another and another of the wires was similarly treated, until escape was easy, when the mouse took his leave, the tourniquet remaining to testify to his shrewdness.

Unaided by this device, a dozen mice could not have bent a single wire."

"Done!" said the fortuitous boot-black, who had been brought in from the street to polish the doctor's vast boots.

"Ah, a very excellent polish!" said the doctor, fumbling in his pocket. "Now, where the mischief —by Jove, I—ah, thanks! thanks!" For the returned miner, noticing his plight, had thrust a dime into the boot-black's hand.

"I really wonder what ailed that ridiculous boy," said Fenton.

"It is indeed most mysterious," said Dr. Binninger. "Were it not that every colored boy instinctively recognizes a Southern gentleman of the old school as his natural friend and protector, I should almost have said that he was terrified by the sight of me. I shall never rest until the mystery is solved."

"'F I see 'im I'll drop y' th' tip," said the head barber. "Next!"

CHAPTER X.

" YES," said Dr. Binninger, " trout fishing is very well for mere amusement ; "but I have seen contests with the scaly folk so much more exciting that I almost wonder at myself for displaying such interest in the day's doings. It is the sporting instinct, gentlemen. I suppose I should kill flies if I couldn't get at lions."

It was mid-April and St. Trout's day. Several of the Travelers had been whipping the brook which ran through the preserves of a country club that counted most of them as members ; and they had ceased to catch fish only to eat fish : and ceased to eat them only to talk about them.

" What was the best fishing you ever saw, Doc?" asked Eckels.

" Hem ! I am not altogether accustomed to the abbreviation of my scientific title, but I will overlook the matter in consideration—"

" No offense intended, doctor," murmured Eckels.

"Ah, quite so. Don't mention it, my dear boy. But to return to the query : the best fishing, or at least, the most plentiful catching of fish, I have ever witnessed was in Manitoba. When mellow autumn gilds the valleys in that far northern clime, the wild grapes, which grow along the banks of the streams, ungathered, fall, over-ripe and fermenting, by tons into the water, transforming every stream for the time being into a fair sample of toddy. It is at these periods that such of the farmers as retain command of their legs are accustomed to pick out dead-drunken fish with their bare hands to salt down for winter, and I can assure you, suh, that fish so cured in the fumes of the gentle juices of the grape, dying as it were in the odor of sanctity, are, even in the salted state, food fit for the gods."

"Was in Ardmore, Pennsylvania, once," said Jim Hart. "Noticed alder bushes along stream all barked and dying. Asked friend why this thus. Carp so thick and eager, jumped out of water after flies, bit bushes, gnawed bark off, bushes dead for miles."

"The German carp is indeed a voracious and hardy fish," said Dr. Binninger. "Recent long and frequent droughts in Western Kansas and Nebraska have proved the German carp to be a fish of peculiar adaptability and versatility. When the streams have dried up, the carp have taken to burrowing in the mud at the bottom to avoid the farmers' boys, who come with baskets to pick 'em up ; but it's hard beating a Kansas farmer. They soon discovered that by plowing the river and lake bottoms, and by running a potato-digging machine along the furrows, they could harvest carp by the wagon load with very little trouble, and the discovery has fairly revolutionized the fish industry of the middle western states.

" The fish," the big doctor went on, leisurely puffing a very strong black cigar, " has often been maligned by so-called scientists who have held him to be a low organization, incapable of arduous thought. Of, late, however, his character has been rescued by other scientists of better-standing and greater accuracy, who, reasoning from wider data, have arrived at a different conclusion. Hem ! perhaps modesty should prevent my speaking in this strain ; hence I will merely relate a little incident of a fisherman who went down into Lake Keuka in ships of the skiff variety in order to fish, trailing a jug of bait behind the boat to keep cool. He was twice annoyed by having his jugs broken, which, of course, put an end to the fishing for the day. On the second occasion when this occurred, the fisherman, soon after, saw a large fish swimming about near the surface in a lazy and irresolute sort of way.

He rowed up to the fish, and, to his surprise, was able to pick it up in his hands. The fish was perfectly sound, but had rather more color than usual, and its breath smelled strongly of alcohol.

"This gave the astute fisherman an idea. Next day he trailed behind his boat a jug, wherein the lure was cunningly commingled with opium. Hid in the stern of the boat, while another rowed, he kept sharp watch. Presently he saw a number of monstrous fish approach, bearing on their noses a stone they had evidently rooted out of the lake bottom. One sharp crack from this broke the jug, when the thieves eagerly drank up the liquid as it mingled with the surrounding water. Shortly afterward a number of the finest fish ever seen on the lake floated on the top in their opium dream of bliss and were captured."

"I think you must be right about the possession of intellect by fish," added Tom Fenton; "and though I lay no claim to rank as a scientist, I can adduce an interesting tale showing the possession by a fish of a high degree of moral rectitude. There have been very many cases where fishes have been caught, in whose capacious maws were found long missing rings, necklaces, baseball masks, and such like trifles. Once in a great while a thieving fish is

stricken by its conscience into a desire to make res-
titution.   Such a fish was the giant cat, which lay
watching the shore of the turbid Kaw all through
the month of April, 1879 or 1880, I am not quite
sure which.   Many fishermen tried to land him, but
he contemptuously refused the most tempting lures,
until a tall man of striking personal appearance came
one day and cast his tackle on the waters.   The big
cat leaped from the water and fell at his feet without waiting for hook and line.   The tall man was
astounded.   On cutting open the fish afterward, he
discovered a gold eagle, which he had lost a year before, lying in the fish's stomach.   Most wonderful
of all, there were besides sixty copper cents, one
year's legal interest, which the noble fish had yielded
up its life in trying to restore with the principal."

   " To return to the more strictly scientific aspects
of the question," said Dr. Binninger, "—not but
that the moral problem is highly interesting—I
should like to speak briefly of the æsthetic rather
than the ethical side of the piscine nature.   Until
some recent discoveries by Prof. Saussier of Vevey,
Switzerland, it was not suspected that fishes were
affected by music.   In a shallow inlet Prof. Saussier found, not long ago, a queer arrangement of
strings in the water which demanded examination.

Viewing the affair from a distance with a water tele-
scope, he saw that some fishes, which had, by the
usual painful method, gained possession of several
fragments of fish-line, were passing them around two
sticks thrust into the water by some fisherman.
When the strings were strung the stakes were
wedged apart by piling stones between them so as
to tighten the strings, three or four fishes rolling a
stone along the bottom with their noses.

" The operation was necessarily slow. The professor watched it at intervals for two or three days.
Finally, when all was ready, the largest fish seized
a stick or bone, and, using it as a plectrum, twanged
the strings with it, while the other fishes gathered
around to hear the music. Of course there was
none, as the submerged strings refused to sound.
After several trials, the fishes tore up their water lyre
in disgust. They had probably caught their idea
from Aimée Saussier, the professor's daughter, who
was in the habit of playing a harp by the bank."

" Speaking of gigantic fish," said Tom Fenton,
" I well remember going up the Ohio in a kick-be-
hind steamer one hot day in August last year. The
river was so low that the sun rose at 11 A. M. and
set at 2. It was like a canyon. At one of the low-
est places I saw an enormous catfish, weighing prob-

ably 500 pounds, lying near a sand-bar which stretched right across the river. A moment after the steamer passed, the wave from the wheel reached the big cat and he floated over the bar. There hadn't been water enough before to float him. Afterward I learned in Pittsburg, that many attempts had been made to capture the fish by wading parties. But whenever a number of men rolled up their trousers and waded in to attempt his discomfiture, the big fish would use his fins to roll himself over and over like a log, and so escape. After two men had had legs broken by trying to stop the rolling fish, the attempt was given up. Once, when the fish was asleep, a steamer struck him and was nearly wrecked. Some of the scales were bumped off his side, and they were nearly as big as my hand."

"And he was never captured?" asked Eckels.

"Never."

"It involves, perhaps, a slight change of the subject," said Eckels, "but an instance in which a fish was discomfited and made to do a signal service to its ancient enemy, the cat, may be in order. Ursula Jenks, of Paducah—in your old state, Dr. Binninger—had a cat which had either outlived or failed to realize its usefulness, and which she threw into a pond, weighted with a six-pound dumb-bell lovingly attached by a string to its neck. Then ensued a strange sight. The cat, with remarkable presence of mind, grasped, as she was going down, at an enormous pickerel, which had been attracted by the prospect of a meal. Away went the pickerel at a frightful speed, the cat held on with the grip of death, and the dumb-bell came plunging after, until the

string, wherewith it had been attached to her, being worn by sharp stones, parted. Then the cat came to the surface and swam ashore."

"By the way, Eckels, my boy," said Dr. Binninger, "how is that little affair of the heart progressing?"

"What affair of what heart?" said Eckels, who

had sat rather gloomily listening to the conversation, but taking no part in it.

"Pshaw, my boy ! Don't attempt to deceive a man of my years and experience, and above all, do not seek to repel the kindly assistance prompted by a sympathetic heart. I presume you have seen the time, suh, when a million dollars wouldn't look so big to you as a single tress of hair of that tint in which the sun loves to linger. Strange thing, this love ! Nothing in the world ever could be so delightful as courtship would be, if it only were. Courtship is the skirmish before the battle of matrimony. Courtship is—when may we congratulate you, sir ? "

"There's no use going around prepared for emergencies," said Eckels, shortly, " because emergencies never happen when you expect them to. I'm not open to congratulations. Congratulate Harry, there. He's big and can stand it."

"Harry ? " said Dr. Binninger. "What nonsense ! Harry is not old enough to marry yet—a mere boy of twenty-six ! It is true that I myself married at twenty-three, but if I may say it, I was very mature and very responsible in business ways. Yes, suh, I— "

"By the way, Dr. Binninger," said Harry, finding

his tongue, "as I am going back very early in the morning and am rather short, would you mind giving me that twenty now?"

"Curious, my dear boy," said the doctor as blandly as if he were conferring an inestimable favor, "but I find that in the hurry of departure, I came up here with only barely sufficient funds myself. But, my dear boy, perhaps some other gentleman—"

"Oh, no matter," said Harry, "I shall manage all right."

The subject of courtship was not again mentioned.

CHAPTER XI.

By one of those quite accidental coincidences
which have to be carefully thought out beforehand,
Harry Porter was striding along Twenty-third street
one afternoon at a rapid pace, regardless of the
throng of shoppers, not a few of whom looked with
pleasure on his tall, strong frame, handsome face
and shining eyes. Ann Copeland was just ahead
of him. She was on the same side of the street, and
going somewhat slower. It is needless to add that
she was presently overtaken.

Apparently Harry's strength was not equal to
further maintaining his rapid pace, for, greeting Ann
with laborious coolness, he walked by her side ;
neither knew whither.

"I am afraid," she said presently, "that I am
wasting your time. You seemed in such a hurry."

"Oh, no," Harry replied. "I have nothing on
hand at the moment except—by the way, did it
ever occur to you that there is no one in this world
who has such a good time as Time ?"

"No—no ; I can't say that it ever did."

"Well, it's true. While the rest of us are hustling

132

around trying to make our living, all Time has to do is just to sit still and elapse. Time neither eats, nor drinks, nor wears rubber overshoes when the

walking is damp, yet the careless creature will out-live all of us."

"Yet time is sometimes weary," said Ann Cope-land, with a sigh.

"Has it seemed so to you?" murmured Harry, bending over her. Then he hastily checked himself and went on : "Time never sleeps, because it never has to. The most it does is to drouse on Sunday afternoons. Time is plain and matter of fact, while Eternity puts on airs and gets its name spelled with a capital letter, but it's all the same old Time. No one knows when Time began being Time, or when it is going to leave off. We are dumped in the middle of it ; that's all we know."

"Why, isn't that the ferry-house ahead?" demanded the girl. "We've walked clear over to the North river. I—I didn't notice. We must go back to the square."

Together they turned to retrace their steps.

"Nan," said Harry softly, plucking at an imaginary fleck of dust upon her sleeve, and watching her cheeks' quick flush, "Nannetta mia, I am something of a liar myself."

"Very possible," she responded with freezing hauteur, suddenly assumed in that way that women have.

"I can tell you more lies in five minutes than Dr. Binninger could in a day, and I'm going to begin now."

"I do not think they will interest me."

"Then you are mistaken. They will interest you, though you may not like them. You are an extremely homely young woman, and I detest you."

Ann looked at him in wide-eyed amazement, as if fearing that he had become suddenly insane.

"You squint," Harry went on with judicial gravity; "your complexion is abominable. While you may fall somewhat short of absolute deformity, you are as ungraceful as a woman can be. I think you dye your hair, and it's a fright at that. I suppose you have the worst taste in dress I ever saw outside of a dime museum, but it really doesn't matter, as no kind of a costume could redeem your innate ugliness. You have a mind to match your appearance. For downright folly, perversity, and all the unamiable attributes that can be in one woman compact, I have never seen your equal. I wouldn't marry you under any circumstances. You are—"

The girl clapped both hands over her face, fully intending to burst into tears. Instead, the humor of the avowal conquered her mood, and she laughed more merrily than she had done for days.

"Do you really love me so much?" she whispered, glancing up into his eyes, then dropping her own. They were walking very slowly now and very close, side by side.

"Indeed I do—dear," he said, "I told you I could lie, but I'd rather not."

For a little while there was silence. A begrimed messenger boy walked backward just before them, grinning up into their faces, but they did not see him. This robbed the performance of interest, and he presently tired of it.

"Tell me some more lies," she said presently.

"Well, you are—"

"No, no. Not that kind," she said, with a sudden tremor in her voice.

"Well, what kind?"

"Oh, I don't care. Fish lies, maybe."

"All right. Once there was a darling girl—no, that isn't the way they begin. Hem! It was owing to the fact that water is an excellent conductor of sound that Jake Watlin, of Sebogatackoconticook Pond, Maine, heard, while fishing there, a curious burring noise, which seemed to proceed from beneath the boat. Presently he caught a large pickerel to which was attached a fine open-faced gold watch, by a black silk cord, jauntily swung about the fish's body, just back of the gills. Mystery of mysteries, the watch was running, and within two minutes and thirty-seven seconds of correct time. Anxious to solve the riddle, Watlin put the fish, just as he was, into a tub of water, transferring him at the dock into a large tank. Close watch was kept on the fish for several days, when he was seen to wind his watch by laying it flat on its back, and turning the stem with his mouth. It was the sound of this operation that Watlin had heard. The watch was presently reclaimed by its owner, who had dropped it into the lake seventeen years before."

"No," said the happy girl; "that isn't right. Tell me something that happened to yourself."

"Well, let me see. Hem! Equal intelligence and far greater fierceness and *savoir faire* were shown by a big catfish in the Ohio river, near Wheeling, whose acquaintance I once made. I was fishing with a party of friends in a boat, when I was suddenly jerked overboard by a gigantic tug at my line. Partly, I suppose, from fright and nervousness, I clung to my line, and was rapidly dragged down into the cool, translucent depths of the water. Here I was horrified at receiving a violent blow on the back of the head. Opening my eyes, I could dimly see a huge catfish, its jaw torn and bleeding, from the hook, holding in its mouth a jagged rock weighing several pounds, with which it was viciously jabbing at my face and head. I dropped the line and rose to the surface exhausted, and was rescued by my friends; a moment after I toppled over in the boat, faint with loss of blood from several frightful gashes. The fish, which escaped, weighed seventy-nine pounds.''

"Oh, you poor boy!"

"Well, dear, you can kiss the scars the naughty fish made, and they'll be well—when we get off this confounded street. By the way, is that the same old ferry house ahead?"

"Why no, it's—I declare, we've walked clear over to the East river!"

They laughed a little, and once more turned back.

"Now I'm going to tell you one or two short lies, that will last just to Fourth avenue," Harry declared; "and then we'll cut across the park. Everything is so pleasant to-day."

"Yes, isn't it?"

"Hem! Once there was a man. Let's see, I guess he lived in Maine and had the mumps. Lots of lies originate in Maine, but I don't know that I ever heard a mump lie. Anyhow, there was a terrible epidemic of mumpery, and a man in Hancock county put his head out of the window. The cold air caused his face to swell very rapidly, and when he tried to draw his head back within the house he could not. He was finally rescued from the outside by means of a fire ladder. This incident seems almost incredible, but it must, in fairness, be said that the window was a very small one."

"Why do you always say 'Hem!' when you begin?"

"I don't know. Custom, I suppose. It's the way the professors do. Here's Third avenue. Now I'll tell a little lie, just one block long. Hem! Kansas has often been called the Sunflower State, a title more than ever appropriate, since the foreman upon Gov. Motley's farm constructed his sunflower clock. Choosing an enormous sunflower, he attached to its drooping head a tiny cornstalk, not more than ten feet long. About the plant he drew on the earth a circle, and divided it into twenty-four parts, each of which was subdivided for minutes and seconds. And now, as the faithful plant, from dawn till dusk, eyes its fierce lord, the cornstalk pointer moves about the dial, indicating the time. The sunflower clock can also be used as a stop-watch, to time races, by holding over it a big umbrella, which checks the revolution upon the instant, when the time, to a fraction of a second, may be read off upon the dial."

They turned northwest, across Madison Square. Once, by the fountain, where the children played, he squeezed her hand.

Then her mood changed little by little, and her doubts began to come back to plague her. The exaltation had been too great to last, and self-questioning succeeded certainty.

"Oh, Harry!" she breathed at last, after a long silence, "I am sure we shall never be happy together!"

"Why not?"

"Oh, I tremble for the future. There was a time when I could have loved you, for I thought you were just a nice, simple, handsome clean-minded gentleman, don't you know. But—but somehow you've changed, or I have learned to know you better. I shrink from linking myself with your glittering intellectuality. I know that women who marry geniuses never are happy, and—"

"Well, did anyone ever see such a goose?" demanded the now thoroughly exasperated Harry.

"Then it was true—all those horrid things you told me!" she demanded, turning upon him like a flash, her big, brown eyes blazing. "A pretty opinion you must have of me! Harry Porter, no man was ever loved as I once loved you; but I will never speak to you again in my life—never ! "

Then she fled up the steps of her own home, and banged the door.

After a minute's perplexity, Harry followed, rang the bell, and asked for her.

The servant, with a grin ill-repressed—for he had witnessed the whole comedy through the sidelight of the hall door—said: " Miss Ann is not at home, sir."

Harry rushed away to Eckels's house, but that worthy also was out.

" He looked as if he wanted to kill somebody," said the waitress and door-maid of the boarding-house, describing Harry's appearance as he asked the whereabouts of his friend and adviser.

CHAPTER XII.

BACK TO THE SOIL.

ECKELS was not at home.

He had accepted an invitation to visit Parker Adams's model farm in Peapack, New Jersey, and had passed the night in the attic of a brown, old farm-house, where the rain of late April pattered in a sudden shower at 4 A.M., rousing him from sleep.

" I have at last discovered," he said to Adams at the breakfast table, " the reason why it is so hard to get up in the morning, and it is one that ought to interest you farmers. It seems that so much American wheat, corn, and gold have been taken to the old continent, that that side of the world has become heavier than ours. The resultant shifting of the earth's centre of gravity further from our side causes an excess of rarefaction in the American air, which, in turn, makes it difficult to undertake severe exertion. Every sailor knows that from Liverpool to New York the voyage is uphill; but the fact has never before been satisfactorily explained."

" Tell that to the Doc. when he comes," said Adams.

"Yes, but, for Heaven's sake, don't call him
' Doc ' ! It's bad enough to be waked at 4, without being killed at 9."

Early in the forenoon Dr. Binninger, Tom Fenton, Jim Hart, the returned miner, and Fraser, the
Canadian, came up and began inspecting the crops.

" Oh, this is very well; very well, indeed, for
Jersey," said Dr. Binninger, at last," as they sat on
the fence by the cornfield, after their labors, smok-
ing; " but nothing to what I have seen. In Gastley
County, Missouri, I once saw the corn growing to
such an unprecedented height, and the stalks so
exceptionally vigorous, that nearly every farmer
stacked up, for winter firewood, great heaps of cornstalks, cut up into cord-wood length by power saws
run by the threshing engines. One man, Barney
Gregory, took advantage of the season to win a fortune by preparing cornstalks for use as telegraph
poles. The *modus operandi* was as follows: Selecting the most promising stalks, Barney removed the
ears while still green, and gave each stalk a daily
injection of dilute potash, made from wood-lye.
This tincture of tree, so to speak, absorbed into the
system, hardened the heart, or marrow, of the corn
until it was quite as tough as cottonwood, and considerably lighter. The stalks, when well painted,

are expected to last twenty years. Of course, they
would not do for city poles, which have to carry a
network of wires, but they are ample for trolley-
posts, and for carrying three or four-wire, country
and suburban telephone and telegraph lines, at a
height of twenty-five feet."

"What is one man's meat is another man's poi-
son," said Fenton. "Fine growing weather, similar
to that which made Gregory's fortune in Missouri,
has come near ruining those of the Western Nebraska farmers who raised pumpkins. Just as, by
all ordinary rules, the crop should have been ready
to house, a mysterious rot began to destroy the
great green globes glowing to yellow in the sun.
An examination by the chemists of the State Agricultural College, showed that the trouble was due
to the too rapid growth of the vines, which dragged
the pumpkins about after them, all over the fields,
until the pumpkins' lower cuticle, being worn out by
the abrasion, they succumbed easily to rot in the
bruised portion. Should another such year come,
the farmers will avoid a like catastrophe by providing each pumpkin with a straw-lined nest, or a little
truck with casters.

"A good illustration of nature's bounty happened
some time ago in Doniphan County, Kansas," contin-

ued Fenton. "A seven-year-old daughter of James Steele was sent, in the middle of the forenoon, to carry a jug of switchel to the men, who were at work near the middle of one of those vast Kansas corn-fields. The corn was about up to little Annie's shoulders as she started, but as she went along it rose and rose before her eyes, shooting out of the soil under the magic influence of the sun and the abundant moisture. Almost crazed with fear, she hastened on, but before she could reach the men, the stalks were waving above her head. The men were threatened in a like manner, but by mounting a little fellow on a big man's shoulders, to act as a look-out, they managed to get out, when they promptly borrowed a dog, to follow little Annie's trail. It was not until late in the afternoon that they reached her, where she lay, having cried herself to sleep, with the tear-stains streaking her plump cheeks."

"The soil of some of the Southern California counties is so rich as to become an actual detriment to the farmer," observed Eckels. "In San Ber-nardino County, a farmer, named Jones, has been forced entirely to abandon the culture of corn, be-cause the stalks, under the influence of the genial sun, mild air, and mellow soil, shoot up into the air so fast that they draw their roots after them; when, of course, the plant dies, as a rule. Cases have been known, however, where cornstalks thus uprooted, and lifted into the air, have survived for some time upon the climate alone."

"Why," said Dr. Binninger, "we used to have the same trouble in Kentucky, but it was solved long ago by burying a heavy stone under each cornstalk, and wiring the stalk down to it. I have known the price of stone to treble in one season in consequence of the purely agricultural demand."

They visited the pig-pen and revived the classic narrative of the fat pig that was put in the bucket and didn't really fill it, though accustomed to eat the bucket full of swill three times daily ; and of the lean pig, whose tail had to be tied in a knot to prevent its crawling through the fence ; they told about the cow that developed such an unnatural appetite for gnawing wood, that she gave, instead of milk, turpentine, shoe-pegs, baseball bats, and bundles of laths.

"And talking of cows," said Fenton, "there is a farmer in Newcastle, Pennsylvania, who had a valuable cow die recently from mysterious causes. A veterinary surgeon made the necessary autopsy and dis·covered in the animal's stomach, a black-snake over four feet long, and a number of small ones, who ran

away so rapidly that only thirty-two could
be killed.   It is supposed
that   the   mother   snake
made her nest in the
animal's stomach by
crawling therein while

the cow was obliged, by a heavy cold
in the head, to sleep with her mouth
open."

"That a cow can fly seems to be beyond the limits of probability," said
Adams, "yet the case seems to be well
authenticated of one of my Jerseys of
light and athletic build, which came
pretty near it.   She jumped her fence
time and again.   After each offense I
built the fence higher, until it had

reached an altitude of twenty-seven feet. At this point I resolved to watch all night, and was rewarded by seeing the animal break off a large number of tree boughs and pile them in a neat heap near the fence. Then, running back a little distance, she came on, pellmell, turning a complete somersault before reaching the heap, and alighting on her back on top of it. The next moment, she was bounced up from the elastic boughs like a jumper from a spring-board, and went over the fence hands down. I have since thought that an occurrence like this might have suggested the story of the cow that jumped over the moon."

They watched the farm hands at work, while Dr. Binninger was reminded of rats.

"The Iowa rats, suh," he said, "are most extra-ordinary. The farmers are accustomed to lock their grain in iron safes, but the rats circumvent even this precaution by gnawing in gangs of three, gang number one refiling number two's teeth, while number three gnaws. These rats sell readily at thir-ty-seven cents each to well drivers, who use the teeth for pointing their diamond drills. It is calcu-lated that the rat-teeth drills do 16 per cent. better work than the ordinary variety."

One of the hands at work with his hoe, a light

mulatto of indescribably ragged raiment, had hoed
up near enough to the group to hear the last words
of the speaker. All at once dropping on his knees,
he raised his clasped hands toward Dr. Binninger
and revealed under his broad hat the fear-palsied
features of Henry the new waiter, of Henry the
runaway boy from the barber shop, of Henry the
latest addition to the force of the Peapack farm.

"Oh, forgive me, Massa!" he moaned, "lemme
go dis time 'n' I'll go furder away w'ere yo'll neber
see me no mo'. 'Deed I will, boss."

"What do you mean?" demanded Adams, leaping
off the fence and striding up to the frightened
youth.

"Oh, don' let 'im hit me, boss! I dun g'way.
I didn't know dishuh farm belong to a Trab'ler
man."

"What is the trouble, Henry?" asked Eckels in
his kindliest tones. "Come, my boy, no one is going to touch you, so long as you behave yourself.
Now what is it?"

"W'y suh, w'en I was in de club, I arsk Mist'
Binninger wha's rest ob story 'bout de ossifer pig,
an' he tol' me, 'n' 'en he said any one heerd one his
stories 'n' didn' belieb' 'em, he was goin' to kill dat
man, sho' nuff, like an' ol' school Southern gemman.

'N' he come to the barber shop 'n' 'e tell stories 'bout wigs 'n' 'possums ; 'n' 'e come yere 'n' tell stories 'bout rats. 'Tain't my fault, Mist' Adams. Ise belieb 'mos' anyt'in ; but yo' des keep dat man 'way from me. Da's all.''

" Henry ! said Dr. Binninger—and his rich, full voice was as soft as a woman's, and all his pompous manner gone ; " you needn't believe you see daylight if you don't want to. As for me, I'm the d—dest old liar that ever left Kentucky. Henry, a true Southern gentleman of the old school is the best friend a nigger ever had. Gentlemen," addressing the rest, "leave me here with Henry, and I reckon we can arrange matters to live on the same earth hereafter."

" Good-hearted old duck, that Binninger," said John Hart. " Say, won't any one but the Doc get anything to eat when the boy is back in the club.''

While sitting on the porch, waiting for the doctor, Tom Fenton told about the success of Mrs. Eph. Dorgan in the poultry business, which, as the reader will remember, was so long the wonder of the people of Pamphila, Delaware, where she lives. Her supply of tender poultry for the Wilmington market seemed to be inexhaustible, yet she never had any lack of remaining hens to lay eggs. It won't be so in fu ture. The neighbors have got on to Mrs. Dorgan's trained hen-hawks. For some time she has had a number of these in her service, ranging the country round, and picking up stray chickens to bring home to her. Of course the losers never suspected human agency, and could have no knowledge that the hawks merely brought their prey to the Dorgan coops. Every night, Mrs. Dorgan used to play "Work While the Day is Dawning," to the hawks on the cottage melodeon, and it was this curious custom which led to the discovery of her nefarious business.

"By the way," said Adams, "see those potatoes over there? They're animal and alphabet potatoes."

"Animal potatoes?"

"Yes. A rare variety. The story of their development is very curious, and is known to but few. The 'animal crackers,' which are sold in groceries to the delight of children, suggested the idea to Elmer Griffin of Palaeopolis, Indiana. Mr. Griffin's four-year-old son was just learning his letters at potato planting time last spring, so Griffin cut with care twenty-six different forms of potato chips, each resembling a letter of the alphabet, besides others in the rude outlines of various animals Last fall he harvested his first crop of alphabet and animal potatoes ; they delighted Griffin's boy greatly, and he is accustomed before eating his potatoes to spell with them simple words, such as 'a·t—at' 'c-a-t—cat' and the like, picking out, at the same time, the animal represented. Griffin has sold $3,167.15 worth of alphabet and animal potatoes at a fancy price, just for seed, and I have the finest patch of them in New Jersey."

" Man modifies nature," said Fenton, " all along the line. There is a spider in Brunswick, Maine, which can write. It is supposed to be in love, and has woven into its web the initials ' W. K.' and ' W. H.' surmounted by a bleeding heart transfixed with Cupid's arrow and a dove. By the way, speaking of the tender passion, what's became of Harry Porter ? Why didn't he come up to-day ? "

"I heard that he was around last night with blood in his eye, looking for Eckels. Speak up, Jack; what have you been doing to turn the path of true love into a corduroy road ? "

" Oh, Lord !" groaned Eckels. " I don't feel very well to-day ! I—I have troubles of my own. I have a bad cough. I—what can be the matter with the young idiot now ? "

CHAPTER XIII.

John Eckels was staring moodily into the fire in the big assembly-room of the Travelers' Club. He was already attired in the festive garb of evening, although it was not yet 6. It was the date fixed for the quarterly Ladies' Night at the club, and Eckels had not spirit enough to take any unnecessary steps, between up and down town.

The party from Peapack came in, noisy and jubilant, discussing the insect pests which make a farmer's life unhappy.

" This is the season when tales of cutworms and other agricultural trials are in order," said Tom Fenton. "And never before this year have such pests been so numerous or so big." A Kentucky farmer, the other day, dug up a bushel of dirt to ascertain precisely how many cutworms there were in it. Carefully sifting the dirt and laying it to one side, while the cutworms were measured by themselves, he presently ascertained that he had just half a bushel of clear dirt left, and rather over a bushel and a quarter of cutworms. This seems a surprising result considering that he had only a

bushel of both to start with ; but the sifting operation took time enough to permit the worms to grow
from a half bushel to one bushel and a half.

"Among worms," Fenton went on, "which have
no conscience, there is no law but the survival of
the fittest. Nevertheless, there is a rude sort of honor
even among grasshoppers, as any one would agree
who had witnessed the extraordinary grass-hopper
tournament at Napa, California, recently. Upon
a bare strip of ground, from which everything green
had been eaten, a hundred grasshoppers tugged
along a single luscious cabbage leaf. Presently one
of them drew a long mark in the dirt, and from this,
in succession, the grasshoppers jumped, the mark
where each landed being kept with scrupulous care.
Three trials were allowed to each contestant, after
which the victor began munching the cabbage leaf,
while the others retired with derisive 'oh's' and
'ah's !' The naturalist who observed this curious
scene measured the winning jump, which was 7 feet
11½ inches. California grasshoppers are very athletic."

"Hem !" broke in Dr. Binninger : "the shrewd
and thrifty air of Connecticut naturally encourages
intellect, even in the animal kingdom. In other
states the potato-bugs sit around on the hills, wait-

ing for the plants to come up. In Connecticut they evade the farmer and his paris-green in a very clever way. Near Moosup, a colony of potato-bugs this year took possession of a waste bit of land and planted their own potatoes. The seed they stole from a farmer who left a bag of seed potatoes in his field over night, gnawing each potato into several pieces, so that a gang of eight bugs and a fore- man could roll it away. The patch they planted was about half an acre in extent, and the bugs kept it nicely weeded, until the plants began to appear above the ground, and then they were in high feather, un- til their fat pickings were accidentally discovered by a wandering mill hand. Then the owner of the land, unfeelingly, appropriated the potatoes, which are es- timated to yield 500 bushels to the acre, and poi- soned off the industrious bugs."

"There are many wonderful things to be seen in the far West," said Donald Fraser, who had just been elected a member of the Travelers'. "In the wonderful climate of California not only plants, but animals, attain a size and vigor elsewhere un- equaled. A peculiar instance of this Brobdingna- gian tendency is found in the giant jack-rabbits of San Mateo. These rabbits have, even within the memory of white men, increased immensely in size

until they now average the dimensions of a Newfoundland dog. With their greater size they de-
velop considerable fierceness, and often attack, in
the open field, men who are seeking to deprive them
of food, by harvesting the cabbages. Comminuted
fractures of the tibia, fibula and femur are common
results of a farm-hand's encounter with a jack-rabbit. The rabbit's only means of defense is to kick
with his great hind feet, but this is by no means to
be despised."

"I like always to consider insect life from the
point of view of the scientist," said Dr. Binninger.
"Bees and ants are especially interesting subjects
of entomological study. At the best, bees have
never been noted for chronological accuracy, though
very industrious and masters of isometry, equian-
gulation and architectonics. It was to the ant that
the sluggard was advised to go for an object lesson
in wisdom. Particularly, no doubt, to the so-called
'farmer ants' of northern Mexico, who are distin-
guished among all their tribe by their passionate
devotion to tilling the soil. These ants have developed tiny cereals growing on stalks a hundredth
part of an inch tall. They dig and level patches of
ground, they sow the grain carefully, saved from
last year's seed, they watch, water and tend it.

When the fields are white with harvest, they nibble down the stalks, place the grain heads in broad leaves, and thresh out the kernels by pounding them with grains of sand. The grain is then stored away in underground granaries for the winter. By thus providing a variety in their diet from the monotonous mule, mining inspector and jack-rabbit in a state of decomposition, these ants have become much larger and stronger, as well as wiser, than those of more northern climes."

" The whole West is full of marvels," said Fraser, " which are unsuspected by the staid dwellers in the prosy East. There is, for instance, a peculiar variety of snake, living near Las Vegas, New Mexico, known by the Spanish-American residents of the territory, as the flagellantes, from a strange custom it has of doing penance for misdeeds. When one of these snakes has, in unrighteous anger, bitten another, he is called into the presence of a solemn jury of his peers, and, so far as naturalists can see, receives a fair trial. If no justification appears, the leading snake—comparable either to the judge or foreman of the jury, or sheriff—hands the culprit a branch of the green mesquit, ' horrid with thorns,' as old Virgil would say. Grasping this frightful weapon in its mouth, the guilty snake flogs its own bleeding sides until the whip falls from its nerveless jaws. Then the surgeon snakes roll the expiating sufferer up in green leaves moistened with spittle, which soon heals the wounds by excluding the air. This ceremony has been witnessed by more than one old hunter, and has caused this variety of snake to be held in superstitious regard."

"The East has wonders quite as great," said Parker Adams. "Fond mothers of Wallingford, Connecticut, desiring to impress upon their male offspring the lesson that one should not bolt his food, are now enforcing it by the sad case of a black-snake, in the Stony Hill region, which recently ran upon a nest of hen's eggs, while in a terribly hungry condition. The famished snake bolted the eggs whole, not knowing that they had been brooded upon by their fond mother until they were near the hatching point. The owner of the hen, attracted by the fowl's angry screaming, came up presently, and found the snake writhing in convulsions, with an expression upon its face, however, that was rather of laughter than of pain. As the man came up, the snake, with a last convulsive wriggle, gave up the ghost. An immediate autopsy was held, by the aid of an ax, and eleven downy little chicks rolled, peeping, out of the snake's interior. They had hatched out in that warm retreat, and by scratching the snake's stomach, had tickled it so that it expired in a fit of Homeric laughter."

"There are escapes as wonderful as that snake's demise," said Fraser. "If you'll pardon another reference to the West, I should like to relate a little incident in the life of a western dog. Bob Jones of Colcannon, Colorado, has a climbing dog, which is, probably, the only living creature that has ever fallen a thousand feet and emerged alive. Scamp—that's the dog—was gamboling about on the edge of the Grand Canyon, when it saw a butterfly, and, in snapping at the airy, flitting creature, lost its footing, and went down kerplulp. By devious ways, and with the expenditure of several hours' time, Jones reached the bottom of the canyon, when, after a long search, he was rewarded by hearing a plaintive whine from the bottom of a deep 'well-hole,' drilled in ages past by whirling and swirling water. Here was Master Scamp, alive, but subdued in spirit. He owed his escape to the sudden descent into the hole, which he exactly fitted, thus forming underneath him a cushion of compressed air. This principle, it is well known, is used in the mechanism, which catches the recoil of cannon, and in the cup at the bottom of the elevator shaft.

"Another Western wonder, I might add. Milk farmers are often accused of milking the pump-handle, but Milt Wiltshire of Gadena, Nebraska, openly avows that he does so, and is blamed by none. Upon Wiltshire's farm there has been discovered a subterranean vein of perfect mineral milk, responding thoroughly to every chemical and gustatorial test. Babies thrive so upon the milk that he is able to charge an extra price per quart, and as the supply is large, Wiltshire has in his well 'the potentiality of wealth beyond the dreams of avarice,' as Dr. Johnson said of Mrs. Thrale's beer vats."

"It has often puzzled me to think what a plain farmer, suddenly enriched, could find to do with so much money," said Adams.

"I understand," said Fraser, "that Wiltshire intends to start a new monthly magazine."

"Oh, in that case——"

"The relation of the wonders of the West could never be exhausted," said Fraser; "but if you'll pardon me the addition of a single instance, I will tell you of the new stand bought by Mrs. Brown of Defiance, Ohio. It came from Grand Rapids, where they make furniture pretty quickly, out of green wood. She set the stand in a big window, where it was exposed to the direct rays of the sun, and pres-

ently it sprouted and began to grow. Noticing this,
Mrs. Brown placed each leg of the stand in a pot of
water, and watered it assiduously every day; and
she has now a fine dining-table, with extra leaves to
insert when she has company.''

"Eckels ! oh, Eckels !" shouted Adams ; "wake
up and contribute something to the general fund
of information. Rise, and give your experience,
brother !"

A wan and dejected countenance looked up from
the corner by the big fire, and fell again; the chin
upon the broad expanse of shirt-bosom, its only
support. He looked up again, when Harry Porter
came banging into the room.

"I want to see you, Eckels," he said; and there
was a lurking devil in his sneer.

"Now little Jacky's going to catch it," said Adams, comfortably nestling into the big chair Eckels
left with a sigh.

And he did.

"You're a pretty adviser !" Harry began, when
the two had vanished from the busier scene.

"Yes," said Eckels, "I am."

"I tried your scheme faithfully, didn't I ?"

"Yes, you did."

"Well, I look as if it had succeeded, don't I ?"

" No, I can't say that you do."

" That's strange ! Yes, it succeeded ! Oh, yes ! She said I was too beastly intellectual to tie to. That's what a fellow gets by straining his intellect to tell lies and extemporize painfully learned epigrams."

" Strange, how a woman can misjudge a man ! "

" Yes, isn't it ? By the way, how is your own business coming on ? Not that I care a d——."

" Oh, about like yours."

" Singular ! This wise man that knows all about the female sex can't run a courtship on his own plans and specifications. I think I was a blamed idiot to listen to you."

" You were, my boy. See here, Harry, you feel sore, don't you ? Well, go ahead ! Keep on kicking ! It relieves your feelings, and doesn't hurt mine a bit. I'm in the dumps so far that you can't drive me any deeper by jumping on me. In fact, it relieves my misery by imparting to my misspent life a taste of variety. Go ahead. You help make things a little more endurable."

" Why, Jack," said Harry, " what's the matter ? Tell me all about it ! I don't blame you for what has happened. I presume your plan was better than my execution of it. Anyhow, that's past. What went wrong in your case ? "

" Oh, not much. She simply opened her eyes very wide, and said, she'd never considered me in the light of a suitor. I overdid the studied neglect act, just as you overdid the intellectual lay. The trouble with me is, I'm too confounded smart. And now I get it in the neck. Serves me right."

" True, oh, philosoph. Are the sisters going to be here to-night ? "

" I don't know. Go ask Langdon."

CHAPTER XIV.

LA BELLE DAME JAMAIS SANS MERCI.

THE Ladies' Nights of the Travelers' Club were such enjoyable occasions that no one need suppose that Miss Copeland and Miss Ann Copeland attended, on the night in question, in the hope of seeing Eckels and Harry Porter. Indeed they had no particular reason to expect either gentleman to be present. Probably they assured themselves that they hoped Eckels and Harry would have the good taste to keep away. In their hearts—but who knows the heart of woman? Certainly not women.

Mrs. Langdon had not seen Dr. Binninger since he dined at her house, and the sight of his expansive form, swelling with pride as he officiated as chairman of the reception committee, reminded her of a remark he had made on the former occasion. So when the enlarged party was gathered in the big assembly-room for the story telling, with which Ladies' Night at the Travelers' always closed, after the early birds had flown, and the chairs were drawn about the fire, she said: " Dr. Binninger, what is a 'sidewinder'?"

" Bless me, I don't know."

"Why, you must. It was you, who said they

were among the most interesting of American snakes, and then you forgot to tell about them."

"Oh, yes; the sidewinders. Why, of course! Bless me, my dear Madame! I'd forgotten all about them. I must be getting old and forgetful. Sidewinders? Why, naturally! Hem! The horned rattlesnake, or 'sidewinder' of California," he went on after sparring for a time until he collected his thoughts, "is a strange creature, named for its peculiar gait. Its peculiarity of locomotion came to my notice in this manner: Gid Marsh has a bonanza farm in Tuolomne county, and during the season of 1879, some of his men killed some of these sidewinders. The relatives of the dead snakes, after attending to the obsequies, ranged themselves in a line at one side of Marsh's immense wheat-field, and, at the signal of command from the oldest and largest sidewinder, they swept like a vast scythe across the field, leveling every blade as if a storm had swept them. The damage was estimated at $579,-813.71. Since then I have noticed that the sidewinder invariably travels laterally, not longitudinally.

"In the face of such singular instances of concerted action," Dr. Binninger went on, "who can deny the essential identity between what we call in animals, instinct and in ourselves, reason? So many instances are known of the imitative tendencies of animals that one sometimes wonders why they have chosen examples so vile. Bob Jones, a prominent saloon-keeper whom I used to know in Louisville, has a dog, for instance, which is not only fond of beer, but has taught his chums the same taste, for a purpose which now appears. One day a tin pail full of beer was given to Spot, who carefully lugged it away to the woodshed, instead of lapping it up at once. Investigation revealed the fact that he had placed the tin on a board raised upon two blocks to form a bar. Ranged on the other side were half a dozen dogs, who had come with bones, rags and bits of biscuit to buy a drink. So many laps were allowed in return for each article, and the other dogs kept coming. By and by Spot, seeing his tin nearly empty, dashed back into the saloon with it, dipped it full of water from a pail on the floor and returned. But the first dog that tasted of the diluted beverage, raised a howl of protest, whereupon the customers attacked Spot in a body, and not only thrashed him roundly, but took away all his riches."

A dusky form glided in front of the circle to cast fresh logs upon the waning coals, and the gleam of the white teeth in Henry's smile as he looked upon the doctor, showed that he had no more fears of the

consequences of disbelief.   As Dr. Binninger laid a
caressing hand upon the boy's shoulder, he fairly
beamed with happiness.

 " That's a good boy, Henry," said the burly narrator; "keep a good fire. We're not ready to go
yet.   This reminds me of the great fires we had over
in the Swiss mountains last winter.   Some of you
may not have noticed that the good old American
sport of sliding down hill in the 'belly-buster' style,
beloved of the small boy, has been introduced into
that country. I am, myself, an honorary member of
a daring club of Italian mountaineers, who have established a toboggan course from the Matterhorn
peak southeastward. It was from the Matterhorn that
Hadow and Croz lost their lives in a straightaway
fall of 4,000 feet when the mountain was first scaled
by Whymper.   On the Italian side the jump is only
about 2,700 feet, which is taken flying; the tobog-
gans are provided with a complicated arrangement of
springs to break the fall, and life is sustained during
the flight, through space, by sucking pure oxygen
through a tube from a magazine under the seat.
The tremendous speed gained by the toboggan during the descent and on the slope below, carries the
sportsman right on and over the gently rounded
summit of Monte Rosa.   When ladies take the slide

they generally use sleds provided with parachutes, which make the precipitous part of the descent safe and pleasant. To get back to the starting point involves a railroad detour of 117 miles and two days' climbing from Zermatt, so that one can't easily enjoy the experience oftener than once a week."

"Can't you tell some stories to match that, Mr. Porter?" said Miss Curtis, the pale girl in blue, who had come with the wife of an elderly member.

She smiled her sweetest smile as she said it, but Harry groaned inwardly, and Ann Copeland ground her teeth in ineffectual rage, not so much at the request as at the smile.

"I can," said Harry, manfully, resisting an instant's inclination to run away, "but not even to oblige you will I. I must beg to be excused. With me such a story as you've just heard means a half day's struggle with a pen, a painful effort to commit the thing to memory and all that. So you see I can't do anything extempore like this; but I'm a fine listener. So I'll beg to remain in the background."

In retiring to the background, Harry by the merest accident had taken a chair, which stood next to that occupied by Ann Copeland.

"Yes, it's all true, Nannette," he soon found occasion to say to her under cover of the laughter evoked by one of Adams's best tales. "I'm just such a fool, Lord help me! Can't we—can't we patch up things between us somehow?"

"And you took all that trouble to please me?" she murmured. "I—I've completely forgotten what I said to you the other night."

"So have I, dear; I fancy you never said it at all."

"I know I didn't."

Fraser was relating the wonderful escape of a bicycle policeman, and had reached that point in the narrative where the pursued scorcher dropped his wheel, and ran up the stairway of a tenement.

"The policeman followed right upstairs, without dismounting, until he came to the stairs leading to the roof scuttle," he went on. "While he was climbing these, and dragging the bicycle after him, the thief got a good start. On the flat roof the policeman mounted again, and overhauled his man just at the end of the block. Such was their impetus, however, that both went over the parapet to the street, a distance of fifty feet, but as they fell on top of the bicycle, the pneumatic tires saved them both from injury."

"But, after all," said Dr. Binninger, "that was nothing to a narrow escape I once witnessed down in the Indian Territory. An Indian woman, while at work, had left her baby lying on a blanket. As I was riding by, an eagle swooped down out of the sky, and, grasping the pappoose in his iron talons, rapidly flew away. I stood aghast. I had no

weapon save a revolver, whose range was too short for the work. Besides, I was afraid of hitting the child. However, up dashed an Indian brave on the run, and hastily aiming his rifle at the flying bird, fired twice, and down it came with a—"

" But wasn't the child smashed in falling ? " asked Nina Markham, wide-eyed.

" Down came the eagle slowly, flapping desperately with his crippled wings. The Indian had broken each with a shot, so that the eagle fell, indeed, but very slowly, struggling to the earth. The child was only slightly shaken up."

" And yet Cooper says that Indians are not as good shots as white men," said Mrs. Langdon.

" And they are not. I neglected to say that the marksman was a half-breed. His eyes, arms, and judgment were all white. His legs were the Indian half of him."

" I like the clear glow of this fire," said Langdon, after a pause, during which every one sat seeing castles in the flickering light. " Fire is swift decay. I once knew a peculiar case, which really seemed to be a cross between phosphorescence and combustion, having all the quality of the former and the brilliance of the latter. The phosphorescent effect of fungi upon the old bark of decaying trees is well known ; but it is not so thoroughly understood that there is in California and Nevada a rare tree, known as the 'witch tree,' from the way in which every limb, twig and leaf stands outlined in the darkest night by some peculiar quality of the bark. A large tree of this species is estimated to furnish illumination equivalent to 100 electric lights of standard sixteen-candle power variety, and a few small towns in the foothills have planted witch trees on street corners, as an economical means of municipal illumination."

"Do you believe in luck?" asked Adams of no one in particular.

"I do," said Harry Porter, promptly, and then he blushed and wished he had not spoken.

"I am myself," said Adams, hastily, to cover the young man's confusion, "a thorough believer in signs, omens, and hoodoos. I have seen too many instances to doubt. Many railroad men have such a prejudice against the number thirteen that most roads skip that numeral in numbering cars or engines. There was a No. 13 freight car on the M. K. and Q., however, until the cyclone came which terminated its career. The path of the storm was very narrow, which, of course, only made its violence so much the greater. It came across the track where a long freight train was running, and ripped No. 13 out of the line like a feather. None of the other cars felt more than an ordinarily high wind. Indeed, the absence of No. 13 was not at once noted. In sailing away, it jerked together the cars before and behind it, when the coupling links, which had been melted by the rapidity with which they had been broken, were welded together again as they cooled, leaving the train as safe as ever, but one car short. Some farmers' boys saw the car as it went off to leeward, but its irons and timbers were probably fused to ashes, by impact with the atmosphere ; as no trace of it was ever seen.

" There is good luck as well as bad," he went on. " I suppose you all know that the good old days in South Africa, when a man could make a fortune easily, are gone, never to return. Nowadays the money is made in London by speculators, and there is no opportunity for such coups and coops as laid the foundations of Carney Carnatto's fortune of $1,500,-000,000. Carnatto went into the Witwatersrand region in 1877 with no capital save a dozen hens, a skillet, and a little pepper and salt. With these and other stuff, purchased from time to time as money came in, Carnatto opened the Digger's Delight restaurant. Whenever he killed a tender broiler, of thirteen years or so, Carnatto would deftly remove from its crop the rough diamonds it had swallowed, as an adjuvant to digestion. By the time the twelve hens were all eaten, Carney had cleared up stones valued at about £7,390, 10s. 6d. With this capital he went into the brokerage business, with the result which all the world now knows."

" By the way," said Eckels, " the mention of South Africa reminds me that we really must have the rest of that story about Jameson at Krugersdorp."

There was a vibrant ring in Eckels' voice, and a new elasticity in his manner. His eyes shone as brightly as—well, no one could see Miss Copeland's for comparison, for she kept them modestly lowered. They were sitting close together, a little withdrawn from the light of the blazing fire.

" Yes, yes ; the Krugersdorp story," said half a dozen voices, and all eyes were turned upon a quiet man away at the end of the semi-circle. But there was no reponse.

" Punch him somebody," said Eckels ; " he's dreaming over his thrilling experiences."

" Me ?" said the returned miner, looking up ; " I wasn't with Jameson at Krugersdorp."

THE END.